Towards Global Sustainability

Paul J.J. Welfens • Jens K. Perret •
Tony Irawan • Evgeniya Yushkova

Towards Global Sustainability

Issues, New Indicators and Economic Policy

Springer

Paul J.J. Welfens
European Institute for International Economic Relations (EIIW)
University of Wuppertal
Wuppertal, Germany

AICGS/Johns Hopkins University
Washington, DC
USA

IZA
Bonn, Germany

Sciences Po
Paris, France

Jens K. Perret
European Institute for International Economic Relations (EIIW)
University of Wuppertal
Wuppertal, Germany

Tony Irawan
European Institute for International Economic Relations (EIIW)
University of Wuppertal
Wuppertal, Germany

Evgeniya Yushkova
European Institute for International Economic Relations (EIIW)
University of Wuppertal
Wuppertal, Germany

ISBN 978-3-319-18665-8 ISBN 978-3-319-18666-5 (eBook)
DOI 10.1007/978-3-319-18666-5

Library of Congress Control Number: 2015947510

Springer Cham Heidelberg New York Dordrecht London

Printed on acid-free paper

Springer International Publishing AG Switzerland is part of Springer Science+Business Media (www.springer.com)

Foreword

Sustainability is a key challenge in the twenty-first century, namely for science, policy makers, the business community, and everybody else on the planet. Policy makers are increasingly looking for indicators which describe the level of sustainability achieved by economics. Given the broad range of complex indicators available in the literature, it seems necessary to create a consistent new indicator which is relevant, easy to understand, and in line with the OECD guidelines on composite indicators.

The EIIW-vita Sustainability Indicator (EVSI) looks at a combination of three pillars: the share of renewable energy, the modified savings ratio—according to the World Bank—and the relative export–import position in the field of environmentally friendly goods. This combination provides a novel and interesting view of the world economy as well as of individual countries. In an extended version of the indicator, water productivity has been added as a fourth pillar to reflect the importance that water plays for the survival of mankind.

It is important to analyze both the positioning of some of the key countries such as the USA, the UK, Germany, France, Italy, the BRIICs—including Indonesia— and to take a look at their recent environmental policy approaches.

Combining insights from this new indicator with well-known statistics from the Human Development Index (HDI) sheds further light on the long-term well-being of the population in different countries. As the HDI and the EVSI are complementary, one should indeed also take a closer look at this combined indicator—with the dimensions of per capita income, life expectancy, and education, the HDI provides valuable additional information beyond standard statistics on per capita income.

The "vita Foundation for Environment, Education and Culture" together with the European Institute for International Economic Relations initiated the indicator project in 2009 and since then has continuously supported the research as well as conferences about sustainability in both Wuppertal and Brussels. The objective of the vita Foundation is to support projects committed to public welfare, i.e., preserving and promoting the basis for a good life.

There has been already considerable international interest in the EVSI, and it should be mentioned that the original Welfens–Perret–Erdem paper on the sustainability indicator is also available in Russian and Chinese. The international EIIW team under the leadership of Paul Welfens has continued the pioneering indicator approach and refined it over the years. With ever more data points available, the perspectives on future empirical research are improving. Finally, the indicator is a global indicator in the sense that not only individual countries' rankings and positions can be studied, but indicator values can also be aggregated across countries so that a global sustainability position can be calculated and compared over time.

This book should stimulate the international discussion on sustainability and the use of adequate indicators. Since the banking crisis, the resource constraints for many countries facing increased foreign indebtedness have accentuated, and this makes it all the more important to carefully analyze both the core dimensions of sustainability and the implications of sustainability indicator analysis. From the gradual changes in the indicator ranking, it becomes obvious that improvements will need a comprehensive concept and will take considerable time to be realized. Both policymakers and markets have their roles to play in achieving green progress in the world economy. As a former McCloy Scholar at the Kennedy School of Government at Harvard University, I have always been interested in contributing to social, environmental, and economic progress, and one may hope that this new indicator is a crucial element in a broad international joint venture for a sustainable world economy.

Oberursel, Germany Frank B. Müller
March 2015

Preface

The world economy is growing and at the same time is facing serious problems with regard to sustainability. This challenge exists despite the progress we have witnessed, for example, in Europe and the USA in the field of improving the water quality of rivers and hence increasing the fish population. Global population growth, along with economic growth, brings sustainability challenges; CO_2 problems and global warming, respectively, are well-known dimensions of modern sustainability issues. The internet has created a new perception of the various regional and international dimensions of sustainability: As regards many big countries, there is concern about carbon-intensive long run growth on the one hand, while on the other hand there is a basic level of optimism due to the enormity of the technological progress in so many fields—this includes, for example, Green Information and Communication Technology (ICT). Indeed ICT dynamics are remarkable: There is visible progress in resource-saving, there is a big rise in energy efficiency, and there is a growing digital network of people interested in the core issues of sustainability. With respect to the economic dimension, it is certainly important to develop monitoring tools and early warning signs which can guide the activities of policy makers, investors, and households, respectively: Consistent indicator sets are required in the field of sustainability.

As regards economic growth, it is fairly clear that this planet is facing a difficult challenge stemming from the combination of both global population growth and ongoing economic growth. More emphasis on green growth and on sustainability is needed.

As regards indicators on sustainability, there is a rich variety of indicator systems in the literature, some of which contain more than a 100 individual indicators. A very complex structure of indicators is, however, often confusing, and indicators which are not in line with OECD requirements on composite indicators are also not ideal. The new indicator presented subsequently—based on an article in the Journal *International Economics and Economic Policy*—is both in line with the OECD rules and rather compact. At the same time, it covers, with its three/four pillars, key dimensions of sustainability: We look at the share of renewables in electricity production, the role of the "genuine savings rate" (as defined by the World Bank), and the country's respective specialization in environmentally friendly exports; in a modified approach, we also look at water productivity. The composite EIIW-vita sustainability indicator is available for many countries and for many years, so that it is also a useful database for statistical and empirical research. Here, we present the theoretical basis of this indicator and also explain its policy relevance.

Looking at the sustainability indicator over time, considerable dynamics amongst both the top 20 countries and the bottom 20 countries can be discerned. Hence, countries which adopt adequate reform policies can improve their respective ranking, and this might have a positive echo effect from national and international investors, who are often increasingly interested in countries and firms which have a more long-term strategic horizon. All those who focus seriously on sustainable development will have to adopt a long time horizon, and many investors have learned from the banking crisis that indeed professional, long-term investment is key to lasting economic performance and manageable risk. Since the banking crisis, western countries have learned that cooperation within the G8 is not enough; the broader scale of the G20 countries has become a new policy forum—in addition to the UN and specific UN institutions active in the field of sustainability. As it turns out, the sustainability indicator for some developing countries has improved over time and this suggests that it is not only industrialized countries which can make progress in the field of sustainability. Comparing indicator results across countries and over time should stimulate benchmarking and the transfer of best practice; it should also help to bring a ray of optimism to the sustainability debate. More competition and more cooperation could be useful elements for achieving a more sustainable world economy. As one of the pillars emphasizes international green competitiveness—and thus indirectly the green innovativeness which is the basis for such competitiveness—the new indicator has one particularly interesting Schumpeterian element which could be distinctly useful. It is, of course, not always easy to identify which innovations and products are environment-friendly; however, one should be pragmatic here and consider the respective OECD classification used subsequently as useful.

The vita Foundation, Oberursel (Germany), has supported our indicator research for several years and we express our gratitude for this. With the publication of this book immediately prior to the UN climate conference in Paris, it is hoped that it will serve to stimulate the broader debate about sustainable development worldwide.

We appreciate the technical support and the comments on part of our research of Samir Kadiric and Vladimir Udalov (EIIW/University of Wuppertal). The intellectual exchanges with Peter Bartelmus, Columbia University, and Raimund Bleischwitz, Wuppertal Institute for Climate, Environment and Energy—now University College London—are also appreciated. We are grateful for the editorial assistance of David Hanrahan (EIIW). The usual disclaimer holds for this special project report that is presented on the anniversary of the EIIW which was founded in 1995 in Potsdam.

The EIIW will celebrate its 20th anniversary in 2015. We are proud that many excellent young researchers and renowned colleagues have contributed to our workshops and research projects; the EIIW stands for award-winning research in Economics; at the same time, one may emphasize our ability for international networking—bringing together scientists, policy makers, and leading figures from the business community has been a trait of EIIW research over many years. We have published part of our analysis in various languages, including, but not limited to, Russian, Chinese, English, and French. Before key institutions—for example, the US Senate, the European Parliament, the IMF, the German Parliament—one of the authors (Prof. Welfens) has testified on various fields of economic policy. It is quite important that we contribute to a broader public debate on the issue of sustainability; the indicator presented within is part of the international search for achieving sustainability in a consistent way and a lively intellectual debate can contribute to this. The EIIW welcomes visiting research fellows and cooperation with other internationally oriented institutions in the field of sustainability, innovativeness, and international cooperation.

Wuppertal, Germany; Washington DC, US — Paul J.J. Welfens
Wuppertal, Germany — Jens K. Perret; Tony Irawan; Evgeniya Yushkova
March 2015

Contents

List of Figures

List of Tables

Green Modernization Issues and Sustainability Indicators

1

Climate policy has become a key element of modern economic policy in a world economy facing the challenges of long-run economic growth—associated with a rising use of energy—and the need to limit global warming. A global temperature rise of more than 2° relative to 1850 seems to bring a considerable risk of warming and hence there is a need for enhanced energy efficiency on the demand side and for a rising share of renewables, which, however, cannot be achieved without considerable cost. Innovation is a natural part of the process toward a less CO_2-intensive way of energy generation and it is also necessary for sustainability. The latter is a broad concept and requires that current generations adjust their production and consumption patterns—i.e., lifestyles—in such a way that future generations will have no less favorable prospects for economic prosperity and a healthy life than the current population. In the public debate, as in the discussions in the political system, there is a need to understand the comprehensive challenges of sustainability; to that end, a compact composite indicator for sustainability is needed to signal the current ecological stress and the ecological quality, respectively. Thus, this book is about the relevance of constructing and using a global sustainability indicator (GSI).

While the climate policy challenges for OECD countries and the BRIICS, as well as many other countries, are considerable, one should not overlook the fact that environmental policy in Europe has achieved major progress in certain fields over the past decades. The quality of the air in many EU regions was dismal in the 1960s, but filter technology, incentives to reduce SO_2 emissions and to invest in the greening of production, and higher energy efficiency have contributed to regional improvement of air quality. The quality of many rivers in EU countries has dramatically improved in the period 1990–2015; the river Rhine—affecting people in five European, including four EU, countries (Switzerland, Germany, Luxembourg, France, and the Netherlands)—has witnessed the return of many fish populations for whom it had previously been assumed the river was no longer hospitable. The International Commission for the Protection of the Rhine (ICPR 2014) states on its website that in 1/8 of the 800 km long river, the targeted increase of structural diversity had been achieved by the end of 2012. Along almost the

P.J.J. Welfens et al., *Towards Global Sustainability*,
DOI 10.1007/978-3-319-18666-5_1

whole course of the Rhine, the quality of the water has improved over time and this has been achieved through both national policy measures and policy instruments of the European Union. Environmental progress can be achieved, not least by means of the careful long-term cooperation of policymakers from different countries and adequate incentives for industry and households, respectively. As will be emphasized subsequently, water productivity is one of the key elements of sustainability and one specific version of the sustainability indicator developed looks at water productivity. In some countries, water seems not to be a scarce resource; for example, in rainy Ireland, for decades households were not faced with the metering of water consumption until 2013 when the Irish government adopted new legislation in the context of overcoming the banking crisis.

There has been a lack of sustainability in the world economy and many OECD countries, respectively, for what is likely to be several decades now. To the extent that sustainability has a global dimension—as in the case of both climate policy and fighting global warming—one will expect two key problems:

- Free rider issues, as the incentive for countries to cooperate toward a global public good is rather asymmetric. The argument of newly industrializing countries—such as China and India—is to point out that OECD countries, above all the USA and the EU as old industrialized countries, have contributed to the bulk of the existing climate problems over more than 200 years of industrialization. Hence, the newly industrializing countries want to enjoy several decades of economic catching-up without many conditions attached in the field of climate policy.
- As the problem of global warming is a truly global problem with no relevance for the geographical location of emission sources, the implication is then that OECD countries are expected to cut back on emissions of greenhouse gas emissions. Achieving contributions from developing countries can be expected through the policy instrument of joint implementation in the short run and in the long run through joining the existing emission trading schemes; the European Union has already created such a scheme and it would be wise to extend it to many other countries, provided one has a clear and consistent concept for such a strategy.

In the EU, the trading of emission certificates almost fell victim to the transatlantic banking crisis (Welfens et al. 2014) which brought very high volatility to most financial market prices—hence the price of emission certificates also became volatile. At the same time, a medium-term fall of that price was observed so that market-based incentives for the internalization of negative external effects from GHG emissions were falling over time. Other environmental problems have also been compounded over time and many authors have developed broad indicators on the quality of the environment (see, e.g., EPA 2012; European Commission 2009b). Many of these indicators have a specific focus, while several of them are very complex. Not many can easily be considered as being both environmentally and economically relevant and as well as rather easy to understand for policymakers, investors, and the broader citizenry. With the continuous challenge of greenhouse gas emissions and the global warming problem, respectively, more interest has arisen in the field of composite sustainability indicators.

One major problem is that the main source countries of GHG emissions, namely, China and the USA, are not part of a wider international trading of CO_2 emission certificates: The two biggest source countries of the second decade of the twenty-first century are not part of the most efficient way to reduce greenhouse gas emissions (in the USA, some regional trading takes place). This is indeed rather surprising in the case of the USA, whose governments have always emphasized the wisdom of market-based environmental instruments. This inconsistency on the side of the USA finds a related inconsistency on the side of EU countries, where national policymakers in several countries are not really relying on emissions certificate trading but rather use selective intervention to raise the share of renewables in energy generation. That share is one of the key pillars of the EIIW-vita sustainability indicator which is to be presented and other elements come on top.

The subsequent analysis is based on a research paper published by Deniz Erdem, Jens Perret, and Paul Welfens in the Journal *International Economics and Economic Policy* in 2010, in which was presented a first version of the EIIW-vita sustainability indicator: This indicator is comprised of the pillars renewable energy share in total energy, the "genuine savings rate," and the competitive trade position in environmentally friendly goods where this particular concept of the savings rate is assumed to indicate a key aspect of long-run production sustainability. Judging countries on the basis of these two elements (renewable energy share plus genuine savings rate—as defined by the World Bank) and the relative strength in green exports can be justified on certain grounds which will be explained later. A crucial advantage is that everybody can easily understand the indicator which is in the range between −1 and +1; it is also one of the few indicators in the literature which are compatible with OECD standards on composite indicators.

The traditional discussion about CO_2 emissions and greenhouse gases as a source of global warming has been rather static, namely, in the sense that innovation dynamics have not been considered much. Given the global nature of the climate problem, it is natural to develop a more dynamic Schumpeterian perspective and to emphasize a broader international analysis, which takes innovation dynamics and green international competitiveness into account: We discuss key issues of developing a consistent GSI, which should cover the crucial dimensions of sustainability in a simple and straightforward way. The basic elements presented here concern the genuine savings rates—covering not only the depreciation of capital but of the natural capital as well—the international competitiveness of the respective country in the field of environmental (or "green") goods and the share of renewable energy generation. International benchmarking can thus be encouraged and opportunities emphasized—an approach developed here. The countries covered represent roughly 96 % of world GDP (2013), 91 % of global exports (2013), and 91 % of global CO_2 emissions (2011) according to the data of the World Bank World Development Indicators Database. The USA have suffered from a decline in their performance over the period 2000–2011; Germany has improved its performance, as judged by the new composite indicator whose weights are determined from factor analysis (or in a pragmatic way). The analysis also looks at more recent changes for a new global hybrid ecological and economic indicator, namely, an aggregation of the human development index and the EIIW-vita global sustainability indicator.

Sustainability of economic development and environmental policy are key challenges of the twenty-first century. It was only since the UN Stockholm conference of 1972 that sustainable development and global environmental issues became a major topic on the international and national policy agenda. Global warming has been a key issue for all countries of the world economy. It only came on the agenda in the 1980s when scientists increasingly found evidence and developed simulation models which showed that CO_2 emissions—cumulated since industrialization in the atmosphere—were raising the average global temperature so that glaciers would start melting and polar ice could turn to water, thus raising the global water level considerably. Parallel to this scientific discovery was a globally relevant development significant for the world economy, namely, the economic opening-up and catching-up process of China. It started in 1978 and brought China to the global number one position in exports in 2013; living standards for 1.3 billion people increased along with high economic growth; however, breathing in parts of the country during certain periods became rather difficult as smog phenomena clouded the skies of some cities. The good news is that China is one of the world's strongest innovators and green innovation in production and export could help to cope with the new problems in China and elsewhere. In a Schumpeterian perspective, raising the topic of innovation dynamics, it is obvious that achieving more green innovations—innovations that are environmentally friendly—is crucial for global sustainability.

In Germany, the first Ministry of Environmental Affairs was created in 1972 in the state of Bavaria, which is a state in which a considerable share of both value added and jobs depends on tourism. This already points to an important aspect, namely, that successful environmental policy and long-term economic development can be mutually beneficial.

Environmental progress has been achieved in many OECD countries and government intervention in many forms has contributed to this. The OECD's emphasis on green growth at the beginning of the twenty-first century has contributed to the modernization of environmental policy in many countries. However, many newly industrialized countries have also developed their own strategy for sustainability. Simply increasing output per capita is not sufficient to achieve a decent standard of living for current and future generations and, while international negotiations on climate policy are sometimes very tedious, there is no doubt that many people in the world have come to an understanding that a successful global climate policy is important for everybody on this planet. As the search for sustainability and a successful climate policy is a truly global challenge, nobody should expect that international cooperation will be easy. However, at least some progress has been achieved since the UN Kyoto conference in 1997 and the Rio conference in 1992. More countries have developed some understanding for global warming issues and in China the frequent and widespread problems with air pollution has contributed to a broader discussion about sustainability and green growth. The shocking accident at the Fukushima nuclear reactor in 2012 has encouraged some countries to adopt rather radical reforms (Hennicke and Welfens 2012). Germany will phase out nuclear energy by 2021 and Switzerland about a decade later and Belgium is also expected to phase out nuclear energy. This seems to be a costly decision; however, it only appears to be so, as so far nobody officially counts the risk of nuclear accidents of the type that have happened in Fukushima and Chernobyl, respectively. The required minimum insurance for nuclear energy is so ridiculously low in all industrialized countries that potential future negative external effects are not internalized. The cost of an accident with a nuclear meltdown in a major German reactor was estimated to exceed the value of 1 year of Germany's GDP—however, in Germany, nuclear reactors are only required to get insurance for about 2.5 billion euros, i.e., less than 1/1000th of the cost of a very serious accident. As no private insurance company would cover the risk of a nuclear accident with a meltdown of the core in the reactor, nuclear power generation should play no role in a market economy. The fact that it has almost no CO_2 emissions is not a strong argument in favor of nuclear power generation. The idea that the assets of a nuclear power generation company could help to cover damage cost is inconsistent, as the Fukushima incident clearly has shown; TEPCO—the company which suffered the Fukushima accident—would have gone bankrupt if the Japanese government had not stepped in to finance most of the damages. It was only by coincidence that for several weeks, the winds were favorable around Fukushima in the sense that nuclear clouds and particulates were not blown toward Tokyo and it was also a welcome coincidence that no big ship with American tourists was en route close to Fukushima at the time of the accident, since the damage claims of US citizens would have implied the quick bankruptcy of TEPCO within months. No form of power generation gets larger implicit subsidies from governments in so many countries than nuclear energy. Such subsidization is distorting competition and is certainly impairing the growth of renewable energy (Hennicke and Welfens 2012). Long-term growth in the world economy needs affordable and safe energy. A rising role for renewables in energy generation is to be expected, not least since governments in the EU are also stimulating the expansion of renewables with considerable—explicit—subsidization. The effective subsidization through higher electricity prices for German households has amounted to about 20 billion euros in 2014 while the market value of renewable energy was only about two billion euros.

At the same time, one should not overlook ongoing global economic growth—with medium-term growth rates particularly high in Asia—which represents a serious long-term economic challenge. Can one achieve a decoupling of global growth and the use of energy, that is, can greenhouse gas (CO_2 and other gases) emissions be strongly reduced in the long run? Moreover, can OECD countries—and newly industrializing countries—develop a consistent, timely, and efficient climate policy? How much innovation is needed in key fields of the economy to bring about the necessary structural change for sustainable development and a successful climate policy? And how can big actors such as the USA, Japan, Russia, Brazil, India, the EU, China, and others be motivated to cooperate in an efficient and consistent way? As regards China, one of the authors (Paul Welfens) was involved in a project with colleagues from other German universities/research institutes and the Chinese Academy of Sciences in 2011–2012, where the focus was on green growth, which naturally is a very important topic not only for China but for the whole world economy. Two follow-up projects—financed by Germany's National Science Foundation *DFG*—have started in 2014, and the European Institute for International Economic Relations will support both research activities in line with its established research focus on sustainability over so many years.

The Stern review (Stern 2006) has contributed much to a broader international discussion on how to best combat global warming and CO_2 emissions, respectively. There is rather broad scientific consensus that more than 200 years of industrialization have contributed to global warming and this in turn will cause serious problems in many parts of the world economy: Long-term changes in climate could mean increasing rain, precipitation, and flooding; for other regions of the world economy, a rising surface temperature will make living more difficult, while most people—say in India—can hardly afford to escape this problem through investing in air conditioning. If such air conditioners were to be installed, the CO_2 problem would become even worse since traditional air conditioning is energy intensive. There is, however, hope that the rise of renewable energy generation and indeed a growing share of electricity made from renewables, e.g., solar, wind, biomass, and water, can be achieved by 2030. Technological progress naturally is a big issue in the field of environmental progress and fighting greenhouse gas emissions, respectively. By 2020, the EU wants to achieve a share of renewable energy of 20 % and by 2050 a share as high as 80 %.

The EU, however, was not the world's leading source of CO_2 emissions in 2014; these were China and the USA. The world's climate problem is indeed a global problem, since it is the aggregate greenhouse gas emissions of all countries which matter for the warming problems of the globe. If the average temperature should increase by more than 2 % by 2050, the sea levels will noticeably rise, more heavy storms/tornados and extreme weather periods are expected, and many climate-related casualties and serious damages to the physical infrastructure in dozens of countries could become a reality. Scientists, policymakers, insurance companies, and all people around the globe are concerned by this phenomenon. As regards the main source countries of CO_2 emissions as shown in Table 1.1, China became the global number one in 2010.

In Table 1.2, the dynamics of the EIIW-vita sustainability indicator is shown. The top 20 countries as well as the bottom 20 countries are indicated. The top five countries in terms of change of indicator value in the period 2000–2011 were Angola, China, Guinea-Bissau, Germany, and Lithuania. The five biggest losers were Congo, the USA, Tanzania, Ghana, and Luxembourg. Therefore there is no consistent pattern of a particular group of countries for the "premier league": both OECD countries and some developing countries show strong improvements over time.

Global warming stands for only one of the key international environmental issues. Avoiding global warming is a truly global public good, as from a physical

Table 1.1 Main source countries of CO_2 emissions, 1995, 2000, 2005, 2010–2012

Country name	1995	2000	2005	2010	2011	2012
Canada	490,946	567,738	576,741	554,408	557,290	550,547
China[a]	3,320,285	3,405,180	5,790,017	8,286,892		
Germany	930,857	891,516	861,733	829,402	810,441	821,718
India[a]	920,047	1,186,663	1,411,128	2,008,823		
Japan	1,223,687	1,251,461	1,282,128	1,191,067	1,240,632	1,275,611
Korea, Rep.	384,966	441,134	493,502	594,517	624,042	
Russian Federation	1,580,147	1,476,998	1,531,658	1,602,426	1,648,129	1,656,774
UK	553,702	556,667	561,102	504,998	464,036	483,424
USA	5,416,608	5,963,063	6,103,294	5,712,757	5,583,379	5,375,003

[a]Data for China and India from World Bank (2014) in kt, World Development Indicators, for the rest of the countries—OECD. Stat in kt CO_2 equivalent

Data Source: OECD. Stat and World Bank (2014), World Development Indicators

Table 1.2 The top and bottom 20 countries with regards to absolute change in EIIW-vita GSI between 2000 and 2011

Top 20 countries	Absolute change in indicator value	Bottom 20 countries	Absolute change in indicator value
Angola	0.2198	Congo, Rep.	−0.2946
China	0.1791	USA	−0.1625
Guinea-Bissau	0.1500	Tanzania	−0.1270
Germany	0.1447	Ghana	−0.1189
Lithuania	0.0901	Luxembourg	−0.1181
Azerbaijan	0.0882	Maldives	−0.1075
Denmark	0.0863	Peru	−0.0929
Netherlands	0.0756	Guinea	−0.0921
Zimbabwe	0.0739	Haiti	−0.0906
Togo	0.0720	Cameroon	−0.0899
Armenia	0.0701	Panama	−0.0779
Kenya	0.0673	Vietnam	−0.0743
Niger	0.0634	Honduras	−0.0733
Hungary	0.0613	Ecuador	−0.0709
Tajikistan	0.0609	Uruguay	−0.0675
Algeria	0.0596	St. Vincent and the Grenadines	−0.0635
Sudan	0.0592	Trinidad and Tobago	−0.0631
Korea, Rep.	0.0566	Japan	−0.0616
Vanuatu	0.0560	Fiji	−0.0527
Singapore	0.0524	Egypt, Arab Rep.	−0.0500

Data Source: EIIW calculations based on data of World Bank, World Development Indicators and UN Comtrade, DESA/UNSD perspective, it is irrelevant in which country the CO_2 emissions have been generated—the global amount of cumulated greenhouse gas will bring about a certain average surface temperature; if a critical level is exceeded, which is assumed to be +2 % compared to the nineteenth century, there will be very serious ecological and economic challenges for many parts of the world economy. It should be emphasized that there are also other national and trans-boundary emission problems: for example, the emission of sulfur dioxide is critical for the acid rain which causes damage to physical infrastructure and buildings (this is certainly a problem in the USA, Canada, the EU, and many Asian countries). Regional emission trading schemes for SO_2 is an efficient way to internalize the negative external effects from SO_2 emission and such schemes have been implemented in both the USA and the EU.

The USA and the EU together represent about half the world economy. In 2013, the European Union and the USA started negotiations on a Transatlantic Trade and Investment Partnership (TTIP)—the main focus is on trade liberalization, in particular the reduction of non-tariff barriers whose tariff equivalent for many sectors is close to 20 %. As the EU countries and the USA have different standards in environmental policy, there are critics of TTIP who are afraid that the rather high EU standards could be undermined, as the pressure of weak US standards would lead to a downward regulation spiral; however, other observers (Sierra Club 2013) rather hope that TTIP would help to raise US environmental standards, thus moving closer to the EU standards. Transatlantic trade liberalization is largely expected to stimulate exports, income/production, and economic welfare on both sides of the Atlantic (Francois et al. 2013; Welfens and Irawan 2014). An interesting question concerns the issue of to what extent TTIP will contribute to higher global output and higher CO_2/greenhouse gas emissions? In the subsequent analysis, we will take a look at this issue in an innovative way and present an answer of what a CO_2-neutral TTIP project would require; while it is plausible that a transatlantic free trade area will stimulate competition and therefore innovation dynamics, it is unclear what the minimum technological progress rate would have to be—i.e., how strong green innovativeness should be—if the rise of exports and income are not to be associated with higher global CO_2 emissions. In the context of TTIP, one can clearly understand the crucial role of green innovativeness.

If global warming stems from increasing cumulated CO_2 emissions, it is natural for policymakers to consider policy options for reducing such emissions (CO_2 and other greenhouse gases). Coal-fired power stations are a major problem in this context, since the specific emission of CO_2 is high for coal—hard coal and lignite. It is lower for natural gas burnt in power stations. It is (almost) zero for water energy and other renewables such as solar power and wind energy. The problem with renewable energy is that these are intermittent energy-producing options so that a sophisticated system of coordinating all the various renewable sources has to be developed. Renewable energy without adequate software and computer investment will not work. On a very sunny and windy day in Germany, in 2013 there will be 100 % renewable electricity production, although on average—over the whole year—the share of renewable electricity was close to only 25 % in 2013/2014.

Energy generation and the share of renewables in energy generation thus is one key issue of sustainable global development.

As regards CO_2 emissions, a global emission trading system (ETS) would be adequate, or all countries could impose an adequate Pigou tax which gives an incentive to cut greenhouse gas emissions. However, an ETS is superior to a system of national taxation, as an ETS can be established across borders—see the example of the EU. If the emissions of CO_2 are rising in Germany in the context of an economic boom, there is no need to impose an inflexible German Pigou tax on CO_2 emissions; it is more efficient to impose a CO_2 ETS whereby the EU decides about the principles of national allocation of CO_2 emission permits and, based on the market price, there will be incentives for the least costly reductions of CO_2 to be realized in the individual EU countries (by contrast, a national tax rate is only a signal to the national market participants—and to foreign investors). If the price of CO_2 emission certificates is rising, the incentive for emission-reducing innovations is reinforced, while if technological progress or a decline of output or a change of the output structure brings about a reduction of CO_2 emissions, the pressure to adopt emission-reducing technological progress will be weakened.

In the end, voters and investors are interested in an economic and ecological system which is stable and generates long-term wealth. As sustainable development requires that future generations should have at least similar prospects of economic welfare as current generations, it is obvious that an adequate indicator which signals the degree of sustainability of every country—and the overall world—will have to contain as one pillar the dimension of the percentage of renewable energy. Moreover, if living standards for future generation are to be on a level that is at least as high as that of today's generation, the current capital stock—broadly defined—has to be maintained. Thus, not all output can be consumed and there is indeed a need for a minimum savings rate of private households (and government, also broadly defined to include social security). If the current per capita endowment with physical capital is to be maintained in an economy with a stable population, it will be sufficient that savings be high enough to face capital depreciation. In any case, the savings rate should be positive if the current capital stock is to be maintained in the future; this capital stock in combination with knowledge and workers generates the current output or gross domestic product (GDP). However, there are other types of capital, not just the machinery and equipment usually considered in the system of national accounts (SNA).

If one also includes natural capital in the form of natural resources, one will want to consider a broader savings concept, namely, one that also puts the focus on the exploitation/harvesting of natural resources. With the exception of renewable energy sources, all natural resources will, at some point in the future, be exhausted and the exploration rate of such nonrenewable resources should thus be considered when discussing sustainable development. The World Bank has developed a broader concept of a savings rate and this—including the aspect of exploitation of natural resources—will be considered subsequently. Interestingly, this "true savings rate," which often differs from the standard savings rate of the SNA, could serve as an early warning indicator for the euro crisis: the crises in Ireland,

Spain, Portugal, and Greece were signaled early on by a considerable decline of the true savings ratio. That ratio also includes as a positive element the expenditure of households on education, understood to stand for human capital accumulation.

A third indicator to be considered subsequently is a Schumpeterian green innovation indicator in exports of goods and services. Countries that have specialized in such green exports are considered to be able to solve certain environmental problems themselves and also to contribute to environmental problem-solving abroad.

In a nutshell, our basic sustainability indicator—first presented in a paper by Welfens et al. (2010b)—is aimed at considering all three mentioned aspects at the same time:

- The share of renewables in total energy generation
- The true savings ratio of households
- The relative degree of export specialization in green products

While time series data are not available for very many countries, we additionally consider water productivity—as the use of water refers to a natural resource which is rather scarce in some regions of the world economy. From this perspective, we will present two indicator sets for sustainability:

- A composite indicator based on three inputs (with a method of calculation to be explained later)
- A composite four-dimensional indicator which includes water productivity

The indicators—dubbed the EIIW-vita GSI—will be calculated for almost all countries of the world economy and we will then particularly present a time series for the three-pronged basic GSI plus a rather recent snapshot for the four-dimensional indicator (this will be presented for the BRIICS, selected EU countries, and Japan plus the USA). BRIICS means Brazil, Russia, India, Indonesia, China, and South Africa. The selected EU countries are five big countries, namely, Germany, France, the UK, Italy, and Spain plus one crisis country, Greece (where the euro crisis started in late 2009 and early 2010). We will look at the top 20 countries in terms of indicator performance and a snapshot of the bottom 20 countries will also be presented. Data are available for almost all countries in the period 2000–2011. Every indicator has advantages and disadvantages. We are convinced that the pillars of our two indicators have certain advantages:

- These pillars and indicators, respectively, can be easily understood by everybody and thus the indicator gives a signal which is easy to understand for both most laypeople and policymakers.
- The indicator is calculated in such a way that the global average is zero, so that the individual values for individual countries can be aggregated (with certain country weights): thus, one can derive a signal for the world economy—how sustainable is economic development at the global level.

- The indicator also has the advantage of being consistent with OECD requirements on composite indicators (which many other indicators are not meeting, while many other indicators combine dozens of sub-indicators in a rather opaque way for calculating an aggregate indicator).
- As the EIIW-vita GSI can easily be visualized on a world map showing individual countries, it is not difficult to understand the current state of the world economy in terms of sustainability at both the level of individual countries and the world at large.
- A disadvantage is that the weights used here—equal weights for each pillar of the respective indicator—might be unconvincing for some reason. However, weights can also be taken from factor analysis or the user/policymaker might want to adjust weighting individually.

The Role of Composite Indicators in a Theoretical and Political Perspective

Indicators have played a crucial role in economic policy since the 1970s; hence it is not only GDP that is crucial for policymakers and the general public. GDP is a standard indicator used in all countries of the world; however, it has some strange elements, as certain activities which destroy the natural capital stock are considered as value added. The growth rate of real output (GDP) is a standard indicator for economic development used in almost all UN countries.

A well-known indicator which is broader than GDP—or value added—is the human development indicator (HDI), suggested and calculated by the World Bank. The HDI considers per capita income, life expectancy, and education for individual member countries; the basic idea is that a higher life expectancy will raise lifetime economic welfare—for any given per capita income in society. Education is implicitly assumed to be a value in its own right. The international rankings obtained from HDI figures often differ from international ranking by per capita income.

An alternative for covering economic and ecological well-being is the global ecological footprint (see Fig. 1.1). The ecological footprint indicates how large a resource requirement human beings in various countries are imposing and whether or not the absorptive capacity of the globe is not exceeded. Aggregating national economic footprints across countries gives a global ecological footprint of the world society—and it seems that there is no sustainability so far, namely, in the sense that the absorptive capacity of the world (the earth and its ecosystem) is exceeded, so that there will be no sustainable and stable global development.

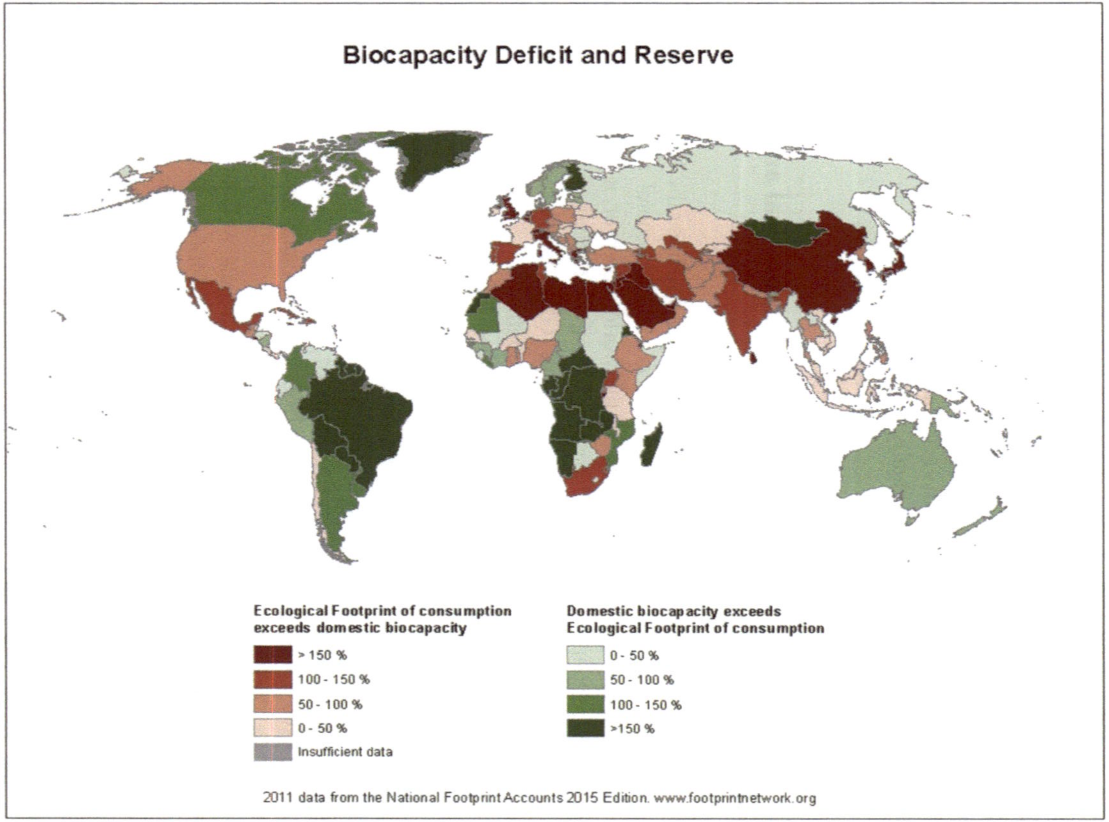

Fig. 1.1 Ecological Wealth of Nations, Results for 2011 from the National Footprint Accounts 2015. *Source*: ©Global Footprint Network (2015) www.footprintnetwork.org

Green Growth and Ecological Modernization

Key sources of greenhouse gas emissions are energy generation, transportation, production of goods and services, and heating. In this perspective, raising the share of renewables in energy generation is crucial and combined modern heating power systems should also be welcome. Sustainability additionally requires that the stock of capital and of natural (nonrenewable) resources be maintained, which raises the issue of the size of the savings rate—with savings defined in an adequate way. Furthermore, one would like to know how many green innovations the economy will generate and how strong the overall technology dynamics translate into a relatively successful specialization in green environmentally friendly export products. These and other aspects play a key role in the EIIW-vita GSI. This indicator is a composite indicator which summarizes several analytical pillars and can give important information to citizens, investors, and policymakers.

The OECD (2010) noted in its interim report on green growth a number of critical observations in the context of the goal to move toward a new growth strategy in OECD countries:

> [...] New indicators and data will be needed to measure progress towards green growth, including to reflect environmental quality, natural resource scarcity and quality-of-life beyond material well-being.
>
> 3. Green growth policies need to be embedded in a coherent, integrated strategy covering demand and supply aspects, both economy-wide and at the sectoral level. This will ensure that green growth is not a just a short-term response to the crisis but a transforming dynamic for both production processes and consumer behaviour. While green growth is relevant to all countries, the policies and approaches used will have to be tailored to specific national circumstances. The overarching priorities for most emerging and developing countries are still poverty eradication, the provision of basic education, ensuring food security, and delivering essential services such as water supply and sanitation. At the same time, a large share of their economies is dependent on natural resources and they are often particularly vulnerable to the impacts of climate change, especially in terms of security of food supply and access to water resources. As such, their economic development will depend on timely adaptation and the sound management of the natural resources that are such a critical base for their economies.
>
> 4. The OECD will deliver a Green Growth Strategy Synthesis Report to the 2011 Ministerial Council Meeting, which will elaborate specific tools and recommendations to help governments to identify the policies that can help achieve the most efficient shift to greener growth. The 2010 Interim Report highlights preliminary findings on a number of key issues that policymakers are currently facing in transitioning to greener economies. These reflect only a sub-set of the broader range of issues that will be addressed in the 2011 Synthesis Report.
>
> 5. The Strategy will develop a framework to help ensure that green growth policies contribute to greater economic integration, technology co-operation and reduced pressure on scarce environmental resources. It will highlight the importance of ensuring that green growth policies are not a source of increasing green protectionism.
>
> 6. Green growth strategies will require a mix of policy instruments, including market-based approaches, regulations and standards, measures to incentivise R&D, and information-based instruments to facilitate consumer choices. Correctly pricing pollution or the exploitation of a scarce resource through taxes, natural resource charges or tradable permit systems should be a central element of the policy mix, most notably to provide a

clear market signal. However, market-based instruments alone will not be enough to bring about a shift to greener consumption and production patterns. Regulations will be needed in cases where market failures result in weak responses to price signals or when a complete ban on certain activities is necessary, for example in the production and use of toxic chemicals. Other approaches, such as voluntary instruments and information-based measures such as energy efficiency ratings and well-designed eco-labelling can play an important supporting role in raising consumer and producer awareness on the environmental impact of specific activities as well as on the availability of clean alternatives.

7. Innovation will be a critical driver of green economies and job creation. Policies to accelerate the development and diffusion of clean technologies and related knowledge will be another key part of the policy mix. As identified in the OECD Innovation Strategy, this will involve a broad approach, comprising price-based instruments and incentives for firms to engage in green activities, as well as public procurement and the funding of basic research. It will be essential to remove barriers to trade in clean technologies as well as to the entry of new firms, and improve the conditions for entrepreneurship, especially in light of growing evidence that young firms represent a large source of more radical innovations. There is also the need for more effective and inclusive multilateral co-operation on science, technology and innovation. The Strategy will address this issue and consider challenges relating to co-operation across countries, funding arrangements, capacity building and international technology transfer. Analysis by the International Energy Agency (IEA), for example, shows that there is considerable potential for the further development and deployment of renewable energy, energy efficiency and other low-carbon technologies. Tapping into this potential will be critical for greening the energy sector.

8. As part of their stimulus packages to respond to the crisis, a number of countries increased public investments in green infrastructure—particularly in terms of public transport, low-carbon energy production, smart electricity grids, energy efficiency of public buildings, and water and sanitation infrastructure. Given that one likely effect of the crisis has been to raise risk premia and therefore lower private investment in higher-risk projects, governments could further build on these measures to move forward investments that would facilitate the development of green technologies and industries. Some countries have also invested in basic R&D to support green innovation and increased their use of environmentally-related taxes. However, not all of the stimulus measures will have been good for the environment, and some may have encouraged investments which could lock in more traditional polluting activities. For example, unless carefully designed, the significant support provided to the automobile industry in some countries, investments in road building and car-scrapping programmes, may have exacerbated pressures on the environment by increasing incentives for private car use.

9. Beyond the crisis, it will be essential to remove policy barriers that hamper the transition to green growth. This involves the reform of environmentally harmful subsidies, the removal of barriers to trade in environmental goods and services, and rationalising conflicting policy instruments. The Interim Report includes a focus on the reform of environmentally-harmful fossil fuel subsidies as an important 'win-win' strategy for green growth and briefly presents some recent developments in greening agricultural support. OECD analysis based on IEA data finds that removing subsidies to fossil fuel consumption in emerging and developing countries could reduce global greenhouse gas emissions by 10 % in 2050 compared with business-as-usual. It would also make these economies more efficient, reduce the burden on government budgets, and alleviate the potentially distortive effects of subsidies on competition.

10. The Interim Report also presents recent OECD analysis on the use of environmentally-related taxes, charges and emission trading schemes. While their use is spreading across OECD and emerging economies, there is considerable scope for expansion in the use of green taxes. Wider use of these market-based instruments can also be an

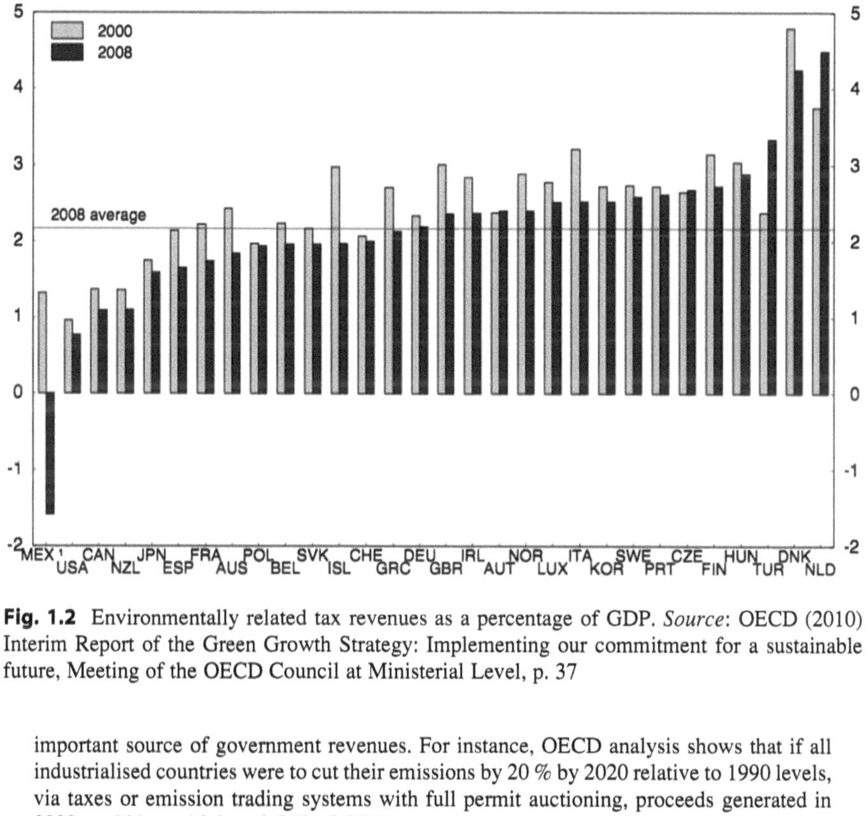

Fig. 1.2 Environmentally related tax revenues as a percentage of GDP. *Source*: OECD (2010) Interim Report of the Green Growth Strategy: Implementing our commitment for a sustainable future, Meeting of the OECD Council at Ministerial Level, p. 37

> important source of government revenues. For instance, OECD analysis shows that if all industrialised countries were to cut their emissions by 20 % by 2020 relative to 1990 levels, via taxes or emission trading systems with full permit auctioning, proceeds generated in 2020 could be as high as 2.5 % of GDP across countries.

It is rather surprising that green tax revenues relative to GDP in OECD countries are in a range from −2 % (Mexico) to 5 % (Denmark, Sweden) as has been shown in the OECD's interim report on green growth; the US share of green taxes was rather low at about 1 % of GDP (see Fig. 1.2). The USA has no national system of CO_2 emission permit trading but ten states have created a joint Regional Greenhouse Gas Initiative (RGGI) as the first mandatory, market-based effort in the USA to cut greenhouse gas emissions. This group of states (Connecticut, Delaware, Maine, Maryland, Massachusetts, New Hampshire, New Jersey, New York, Rhode Island, and Vermont) have capped and will cut CO_2 emissions from the power sector by 10 % in 2018. These states will use the proceeds from the auctioning of CO_2 permits primarily for public benefit and public projects, respectively (OECD 2010).

The share of renewables in total energy has increased in the EU from 8.3 % in 2004 to 14.1 % in 2012. The highest shares of renewables in 2012 were recorded in Sweden (51 %), Latvia (35.8 %), and Finland (34.3 %); Germany's position was in the medium range, namely, 12.4 %. Malta (2.7 %) and Luxembourg (3.1 %) recorded the lowest shares of renewables in 2012. The progress recorded is in line with the EU's goal to raise the share of renewables to 20 % by 2020.

It is not convincing to assume that optimal environmental taxation would bring about such a large range of tax revenue–GDP ratios, namely, between 0 and 5 %. A Pigou tax which helps to reduce emissions and other negative environmental externalities could be useful not only for internalizing negative external effects. Higher tax revenue from environmental taxation is also useful, as it allows to cut other taxes, particularly income taxation. With the exception of Pigou taxes—assuming that they do not exceed the adequate levy for internalizing negative external effect—taxation will cause negative welfare effects. Thus a principle of rational tax policy is obviously to have a full range of environmental taxes implemented. If revenue from taxation is used to stimulate innovation and thus to help internalize positive external effects—that is, to bring about a higher output than pure market forces would bring about in equilibrium—one can expect further positive welfare effects. Positive external effects from (green or other) innovations occur if there are intrasectoral or intersectoral and national or international spillover effects.

Government paying subsidies for innovations can indeed be justified if there are positive external effects: The social benefits then must exceed the private benefits obtained by the innovating company. If the external benefits are 30 % of the private benefits, the subsidy rate should indeed be 30 %. The promotion of green innovations is of particular interest since there is a double externality to be considered:

- The standard positive innovation externality (already mentioned).
- To the extent that negative emission externalities are reduced through green innovations, there is a double externality—if the reduction of negative emission externalities is equivalent to 20 % of the private innovation benefit, the subsidy rate should thus be $30+20=50$ %.

From the green growth perspective, it is quite interesting to consider the revealed comparative advantage (RCA) of countries in the field of green/environment-friendly products. Countries which have a relatively strong specialization in this field—that is, the sectoral export–import ratio for environment-friendly products is higher than the aggregate export–GDP ratio—will obviously contribute strongly to solving environmental problems abroad, and as countries with a strong RCA typically will have strong green innovation dynamics, one may assume that domestic problem-solving in the field of environmental quality improvement will also be rather strong.

Various OECD countries have made considerable progress with green growth, that is to say reducing greenhouse gas emissions, raising resource efficiency, and contributing to more green patent dynamics. Germany, the Netherlands, the UK, and Denmark have been among the EU countries making considerable progress in reducing greenhouse gas emissions in the first decade of the twenty-first century. Within Denmark, the city of Copenhagen—belonging to a network of green cities in the EU—has shown particular progress in green growth and could become the first carbon-neutral city in the EU by 2025 (see Box 1.1).

Box 1.1 Copenhagen, a Green Haven? [Adapted from Jamet (2012, Box 3)]

While cities account for a large share of GHG emissions because they also represent a large share of GDP and population, they are not always the most important polluters when emissions per capita are considered (Hoornweg et al. 2011). Copenhagen stands out as an example in this respect: In 2005, CO_2 emissions per capita in the municipality of Copenhagen were about half the average country rate. This pattern reflects cities' potential to reduce GHG emissions per capita. For instance, higher population density makes public transport more attractive, limiting the use of cars, and makes it easier and less costly to develop district heating systems (OECD 2011). In contrast, some GHG emissions from agriculture are difficult to reduce, explaining relatively large emissions per capita in rural areas. Suburbanization can also contribute strongly to GHG emissions.

Copenhagen is already a low CO_2-emitting city but plans to do even more and to become the first carbon-neutral capital by 2025. Meanwhile, the city targets to cut CO_2 emissions by 20 % between 2005 and 2015. Copenhagen's strategy rests on plans and policies very similar to national ones but also includes some more ambitious ones:

Seventy-five percent of the emission cut would be achieved in the energy sector by moving it away from fossil fuels. Today, most homes in Copenhagen are connected to a district heating system based on combined heat and power plants and incineration of waste, which has allowed reducing CO_2 emissions significantly but remains largely dependent on fossil fuels. Further emission cuts would require increasing the share of renewables in electricity generation. In particular, the municipality plans to develop co-generation from wind and biomass.

The transport sector would account for 10 % of the cut. This will be achieved by favoring walking and bicycling even more. In 2010, already 35 % of all trips to work or for education in the city of Copenhagen were made by bicycle with this share rising to 50 % of trips for people working and living in Copenhagen. The municipality also plans to improve the quality of public transport and to promote car sharing. Stringent performance standards concerning CO_2 emissions from buses are being gradually introduced and the city is experimenting electric buses and municipal cars. Parking places are limited.

Ten percent of the cut would also be achieved in buildings with particular efforts to increase energy efficiency in municipal buildings.

The remaining 5 % of the cut is expected to be achieved through changes in household and firm behavior encouraged by information and education campaigns and through urban development. By continuing on this path, the municipality expects to reduce CO_2 emissions by 45 % between 2005 and 2025. Complete carbon neutrality would be achieved by investing in more

(continued)

Box 1.1 (continued)
windmills or by reforestation to capture more CO_2. While cities have a key role to play in actions to mitigate climate change, they also need to adapt to the impacts of climate change. As a low-lying city, Copenhagen is potentially exposed to coastal flooding that will increase with climate change. The city has already undertaken a number of actions to adapt to these effects of climate change and has developed an "adaptation plan." OECD estimates suggest that, partly thanks to these actions, the city is not particularly vulnerable to sea level events (Hallegatte et al. 2008). Despite these impressive achievements and objectives, Copenhagen's air quality is not among the best in selected OECD cities. Emissions of particulate matter, which have been shown to have large detrimental effects on health, were still relatively high in 2008 despite past reductions. This partly comes from pollution from diesel cars, wood stoves, and other materials (OECD 2009). These emissions may have fallen further in the recent past with the introduction of "low-emission zones"[1] and policies to limit CO_2 emissions will lead to less emissions of particular matter as a co-benefit (Bollen et al. 2009). Nevertheless further efforts may be required in this area.

Company Reporting Initiatives

At the global level, the UN and other institutions have launched initiatives for companies to report on sustainability—broadly defined. There is a Global Reporting Initiative (GRI) which aims at encouraging big companies, typically multinational companies, to regularly publish sustainability reports. The publication of sustainability reports is considered to be a part of corporate social responsibility (CSR). Within the EU, there also is a new initiative on CSR 2010–2014, namely, for companies with more than 500 employees and net sales exceeding 40 million euros (or a balance sheet volume of more than 20 million euros). There are also new international framework agreements (IFAs; Voss et al. 2008) which are relevant for big companies.

As regards the share of greenhouse gas emission in OECD countries, the sectoral split in 2009 was such that the energy sector dominated with a share of 35 %, followed by road transportation (20 %), industry (19 %), agriculture (8 %), residential (7 %), services (4 %), and others (7 %). Across countries, one finds, of course, some variations. There is, however, no doubt that the energy sector, and thus the share of renewable energy, is of prime importance in all countries. OECD countries

[1] Since 2006, the four largest cities in Denmark are allowed to introduce low-emission zones in which heavy vehicles have to meet some standards in terms of emissions of particulate matter.

Table 1.3 Wealth and per capita wealth by type of capital and income group, 1995 and 2005

Income group	Total wealth (US$ billions)	Per capita wealth (US$)	Intangible capital (%)	Produced capital (%)	Natural capital (%)
1995					
Low income	2447	5290	48	12	41
Lower middle income	33950	11330	45	21	34
Upper middle income	36794	73540	68	17	15
High-income OECD	421641	478445	80	18	2
World	504548	103311	76	18	6
2005					
Low income	3597	6138	57	13	30
Lower middle income	58023	16903	1	24	25
Upper middle income	47183	81354	69	16	15
High-income OECD	551964	588315	81	17	2
World	673593	120475	77	18	5

Data Source: World Bank (2011) The Changing Wealth of Nations, Washington, DC, p. 7

have some programs which are designed to help developing countries to raise the share of renewables in the energy sector.

As regards the structure of capital in the world economy, it is interesting to observe that there are certain patterns observed in rich countries and poor countries, respectively (World Bank 2011). The share of natural capital in overall capital is rather large in poor countries—i.e., in countries with a low level of per capita income, as is shown in the subsequent Table 1.3. The share of natural capital in world capital has dropped slightly from 6 % in 1995 to 5 % in 2005; the share of natural capital in low-income countries has declined from 41 % in 1995 to 30 % in 2005. The overall stock of global natural capital has, however, not declined over time: It was $30,273 billion in 1995 and in 2005 the global natural capital stock stood at $33,680 billion. One may, however, criticize the World Bank view that natural capital in high income countries is not playing a strong role, as is suggested by the subsequent table, for two reasons:

- The value of produced capital is likely to be raised by the availability of natural capital; for example, it is well known that in a hedonic pricing approach, the market value of real estate and other physical assets is raised through a clean environment and the absence of physical hazards (including the risk of flooding related to global warming). Including such imputed indirect valuation effects of natural capital and a stable climate would certainly raise the share of natural capital in total capital.

- There is also a problem of international risk spillovers which could go in a north–south direction as well as in an east–west direction or a west–east direction: the notion that investment decisions can cause negative international spillover effects is relatively novel and to the extent that such spillovers are linked to international capital flows free capital flows are doubtful (Welfens 2010): Consider the well-known problem of carbon leakage which occurs for example in the context of industrial production of OECD countries being relocated to developing countries—the relatively CO_2-intensive parts of the value-added chain are relocated to developing countries where the CO_2 intensity of production will, in the case of certain countries, be higher than in OECD countries; the latter is largely related to the fact that certain developing countries strongly rely on fossil fuels. It is rather unclear how the geographical risk incidence of that type of carbon leakage is. There is a need to conduct more research in this field.

The fact that favorable environmental local/regional quality positively affects the market value of houses is firmly established (see as survey Boyle and Kiel 2001; for an excellent Swiss study, see Din et al. 2001).

With the rising accumulation of physical capital and knowledge, there is an increase in per capita income. At first sight, it seems that the relative economic weight of natural capital is declining; however, one may argue that the role of natural capital is rising as a consumption factor, namely, along with the rise of tourism and the increasing length of vacations. Some countries generate about 10 % of GDP from tourism, for example, Switzerland and Austria; from a growth theoretical perspective, one may argue that real value added in tourism $T = f(K, L, N)$ where K is produced capital, L is labor, and N is natural capital; environmental policy and general economic activities will have an effect both on the real market value of natural capital, including land, and on the depreciation rate of natural capital. There is, however, an analytical problem that part of natural capital (such as rivers or mountains) has no private ownership and therefore has only an implicit asset value. Moreover, one cannot expect market forces alone to bring about an internalization of external effects—as is suggested by the Coase theorem. The latter requires that clear property rights be established on the one hand and that there should be small transactions costs in markets on the other, which is certainly not the case in many fields of natural capital damaging and also not the case in the field of negotiations on global warming. Bretschger (2014) has argued that the costs of environmental negotiations in the context of global warming are fairly high. The above aspects obviously have not been covered in the interesting analysis of Pittel and Lippelt (2014). Global warming is a threat for many people in the world and it is partly destroying natural capital through the melting of ice, incidents of flooding, and more intensive hurricanes. Moreover, it will often destroy houses and infrastructure capital—and indeed there have been large damages caused by global warming and natural disasters as is shown in the subsequent graph which reflects the view of an insurance company. Thus, the interests of many companies and many people are aligned in achieving a higher degree of sustainable development.

As regards the destruction of wealth by natural hazards, an event related to global warming Fig. 1.3 indicates key loss events from an insurance perspective. In the post-Kyoto process, it will be very important to face the global climate challenge on a broad scale: simply focusing on OECD countries would not only imply the restriction of attention to a group of countries, which around 2010 were responsible for less than 50 % of global greenhouse gas emissions; it would also mean to ignore the enormous economic and political potential which could be mobilized within a more global cooperation framework. The Copenhagen summit of 2009 tried, without much success, to set a new agenda for long-term climate policy. The USA and China were two big countries which did not really share the EU and Switzerland's strong concerns about global warming.

Ambitious goals envisaged for the long-term reduction of greenhouse gases will require new efforts in many fields, including innovation policy and energy policy. If one is to achieve these goals, major energy producers, such as the USA, Russia, Indonesia, and the traditional OPEC countries, should be part of broader cooperation efforts, which could focus on sustainability issues within a rather general framework:

- Sustainable development, in the sense that national and global resource efficiency strongly increases over time, so that future generations have opportunities equal to those of present generations, in terms of striving for a high living standard.
- Sustainable investment dynamics in the sense that investment in the energy sector should be long-term—given the nature of the complex extraction and production process in the oil and gas sector and in the renewable sector as well (not to mention atomic energy, where nuclear waste stands for a very long-term challenge); investment dynamics will be rather smooth when both major supply-side disruptions and sharp price shocks can be avoided. The current high volatility of both oil prices and gas prices—with both prices linked to each other through some doubtful formula and international agreements—is largely due to instabilities in financial markets: portfolio investors consider investment in oil and gas—in the respective part of the real sector in some cases, in many cases simply into the relevant financial assets—as one element of a broader portfolio decision process, which puts the focus on a wide range of assets, including natural resources.
- Sustainable financial market development: If one could not achieve more long-term decision horizons in the banking sector and the financial sector, respectively, it would be quite difficult to achieve rather stable, long-term growth (minor cyclical changes are, of course, no problem for the development of the energy sector). With more and more countries facing negative spillovers from the US banking crisis, an increasing number of countries will become interested

Fig. 1.3 Loss events in the world economy from an insurance perspective. *Source*: Munich Re, Geo Risks Research

in more stability in global financial markets. At the same time, one may not omit the fact that emission certificate trading systems established in the EU have created a new financial market niche of their own. With more countries joining international emission trading schemes (ETS approaches), the potential role of financial markets in the world's efforts to cope with climate policy challenges will become more important over time. It may also be noted that stable financial markets are required for financing investment and innovation in the energy sector. From this perspective, overcoming the international banking crisis is of paramount importance; however, the progress achieved within the G20 framework is rather modest—not the least because there is still weak regulation for big banks (for which, the problem of too big to fail is relevant) and because more competition, as well as better risk management, has been hardly achieved in 2009; transparency is still lacking, not the least because the IMF has not yet published the Financial Sector Assessment Program for the USA, which is now overdue for many years. Without more stability in financial markets and banks, there is a considerable risk that the creation of new financial instruments associated with emission trading will simply amount to creating a new field of doubtful speculation activities with massive negative international external effects.

Sustainability has thus far not been a major element of economic policy in most OECD countries and in major oil and gas exporters, although the sustainability policy may be considered to be a key element of long-term economic and ecological modernization; sustainability implies a long-term perspective and such a perspective is typical of the oil and gas industry. The use of fossil fuels is, in turn, of key importance for climate change and sustainable development, respectively—and the use of such primary energy sources causes CO_2 emissions. In contrast to general discussions in the international community, which typically puts the focus on CO_2 emissions per unit of GDP (or per capita), it is adequate to consider CO_2 emissions per unit of GDP at purchasing power parities (PPP); otherwise, there would be a crucial bias in the comparison of CO_2 emission intensities. Comparing CO_2 emissions per unit of real GDP of 2000 and 2010, one can see that in almost all countries considered in Fig. 1.4 and 1.5, the specific CO_2 emissions have reduced. As global CO_2 emissions matter to the world economy (emissions can simply be expressed as (CO_2/real GDP) × real GDP), one can point out that the growth rate of total CO_2 emissions is the sum of the growth rate of specific CO_2 emissions and global output growth. If the growth rate of global output is 3 % and the average shrinkage rate of CO_2/real GDP is less than 3 %, the total worldwide CO_2 emissions will still increase.

The PPP figures look quite different from the emission intensities based on nominal $ GDP per capita data; for example, China's performance on a PPP basis was not much worse than that of Poland in 2000; however, in 2010 Poland's position had strongly improved while that of China's specific emission level has reduced only slightly; Russia, Kazakhstan, and Azerbaijan had very high emission levels per unit of GDP in 2000, but in 2010 the situation has clearly improved

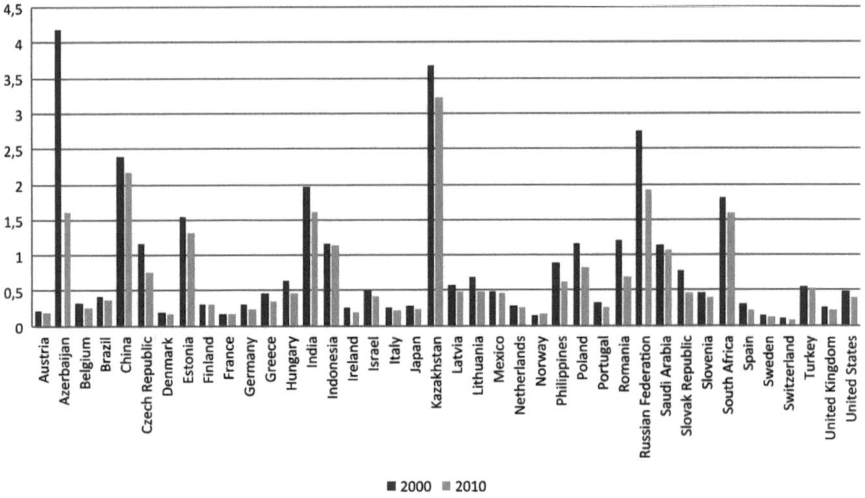

Fig. 1.4 CO_2 emissions (kg per 2005 US$ of GDP). *Data Source*: World Bank (2014), World Development Indicators

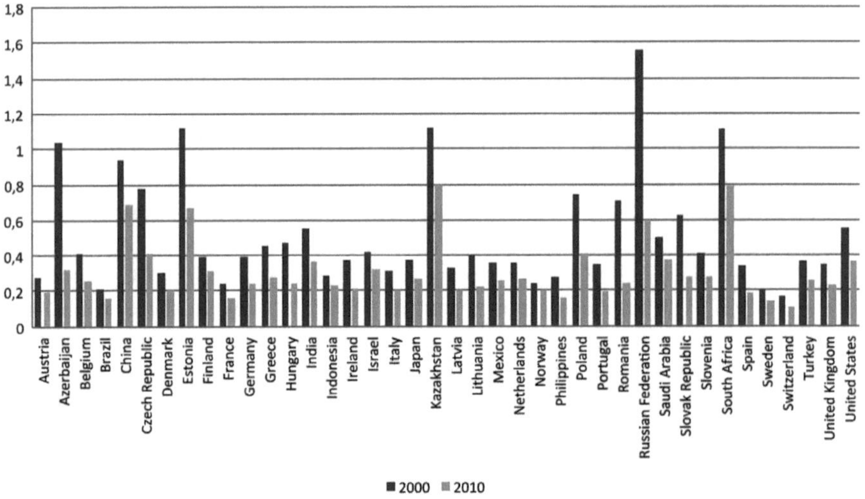

Fig. 1.5 CO_2 emissions (kg per PPP $ of GDP). *Data Source*: World Bank (2014), World Development Indicators

(see Fig. 1.4); looking at kg per PPP $ of GDP, South Africa, Kazakhstan, China, and Estonia have high figures in 2010—very low indicators were observed for Switzerland, Sweden, Brazil, and the Philippines as well as France (see Fig. 1.5). Erdem (2012) has shown that in Eastern European countries, a change in the composition of output in many cases has contributed to a reduction of specific emission levels. In addition to this effect from structural change, one can explain the progress over time through technological modernization. From a comparative international perspective, Switzerland and Sweden are clear leaders in terms of low emission intensity. Austria, France, Italy, Germany, Portugal, and Brazil are also part of the group of leading countries in terms of low greenhouse gas emission intensity.

Greenhouse gas emissions, toxic discharges in industrial production, and deforestation are among the key aspects of global environmental problems. Long-term economic growth in the world economy will intensify certain problems; at the same time, growth is coupled with technological progress, which in turn could allow for a decoupling of economic growth and emissions. It is not clear to what extent countries and companies contribute to solving environmental problems, although some countries—for example, Germany, Switzerland, and Austria—claim that exports in environmental products strongly contribute to overall exports and also to the creation of new jobs (Sprenger 1999). Under the Obama administration, the USA has also taken a fresh approach to environmental policy, where innovation is one key element emphasized in the context of a new approach to climate policy. Given the strong decline of output in light of the transatlantic banking crisis, it is noteworthy that both the US government and governments of EU countries have emphasized incentives for more green investment. However, in the euro crisis, countries facing a major crisis (e.g., Greece, Portugal, Spain, and Ireland) have effectively been forced to strongly cut government expenditures for environmental modernization projects.

While certain fields of environmental problems have seen some improvement over the past decades, for example, the quality of water in many rivers within Europe having improved in the last quarter of the twentieth century, other challenges have not really found a convincing solution. In the EU, the European Environmental Agency (2008) reports on various fields of economic improvement. The BP report (2009) also presents progress in a specific field, namely, the reduction of CO_2 emission per capita in OECD countries; the 2013 report of BP is modestly optimistic that the CO_2 emission intensity in OECD countries—representing more than half of the world economy—will continue to fall in the medium term.

The global picture is different, however. Greenhouse gases have increased over time, and while emission trading in the EU has made considerable progress, the global dynamics of CO_2 and other greenhouse gases have been strong. It is not surprising that Asian countries for example—often with rather low per capita income compared to the USA or the EU—aim at further economic catching-up and high medium-term economic growth. Therefore the demand for energy will rise in Asia, as well as in other areas of the world economy, and rising CO_2 at a global scale could be the result. The governments of some countries, such as Japan, seem to hope that maintaining a high share of nuclear energy generation is a key element for achieving a higher degree of sustainable development. However, while it is true that CO_2 emissions are rather modest in nuclear energy generation, one should not overlook that this form of electricity generation stands for other serious dangers—as has become evident in the Fukushima incident in 2011 and in the explosion of the nuclear reactor in Chernobyl in 1986. The Irena report for 2014 indicates that in that year the share of global investment in renewable energy for the first time has exceeded that share of investment in fossil energy sources.

While global political interest in sustainability issues has increased over time, the recent transatlantic financial market crisis of 2008/2009 has undermined the focus on sustainable development. That crisis has reinforced short-termism in the economic sphere as well as in the political sphere in many OECD countries. Faced with a great recession in the USA and in the EU, political priorities have shifted in favor of restoring economic growth and the focus on the environment has become weaker in many countries.

It is fairly obvious that financial markets shaped by relatively short-term decision horizons—and short-term oriented bonus schemes—are undermining the broader topic of sustainability. It is difficult to embark on more long-term sustainable strategies in companies and households, if both banks and fund managers mainly emphasize short- and medium-term strategies.

In 2008 and for the first time, energy consumption and greenhouse gas emissions were larger outside the OECD than in the OECD countries. This partly reflects the dynamics of successful economic globalization, namely, that countries such as China, India, Indonesia, Brazil, etc., have achieved high, long-term growth, which goes along with rising emissions. Economic globalization has several other aspects, including:

- Enhanced locational competition which reinforces the interest in foreign direct investment and multinational companies
- Higher global economic growth (disregarding here the serious short-term adverse effects of the transatlantic financial crisis and the world recession) which corresponds with stronger competition and a broader international division of labor on the one hand and with potentially fast-rising emissions and growing trade in toxic waste on the other
- The fast growth of transportation services and hence of transportation related emissions which particularly could add to higher CO_2 emissions

From a policy perspective, it is useful to have a comprehensive assessment of the pressure on the environment. Several indicators have been developed in the literature, which give a broader picture of the environmental situation. The EU has emphasized the need to look not only at GDP but at broader measures for measuring progress (European Commission 2009a). In economics, one finds a strong tradition of broader environmental and economic indicators which are supposed to capture the broader picture of welfare beyond the simple figure of GDP or net national income [net national income is gross national income (GNI) minus capital depreciations; the difference between GDP and GNI is the international payments of dividends, interests, and part of the internationally mobile workforce (remittances sent to the home country) plus contributions to international organizations]. The concept of net domestic product (NDP) or net national income is already weakly related to sustainability, as it is obtained as GDP minus capital depreciations and GNI minus capital depreciations, respectively. One should indeed take into account that long-term economic development—sustainable development—is not possible if the physical capital stock is not maintained. Ultimately, one may argue that the main goal of people is a high per capita consumption and a healthy nice life in combination with a high life expectancy, but indeed a clean environment is linked to all three aspects mentioned:

- Environmental hazards and high emissions mean potential health problems for many people, particularly for children and the elderly—a poor environment is not compatible with a healthy life.
- Further global warming and high emissions pose risks to a rise of life expectancy in many countries—including many developing and industrialized countries.

How are we to understand how serious the problem of sustainability is in individual countries and in the global economy? How can we assess progress in sustainability in individual countries and the world economy? Here the focus naturally will be on adequate sustainability indicators which can serve as a signal to the general public, policymakers, and firms. One should also note that an increasing number of firms worldwide have started to publish sustainability reports; those are often broadly defined and sometimes not only focus on environmental sustainability—aspects of CSR are also included by many firms in OECD countries.

Global Sustainability Indicator Approach

Most sustainability indicators are mainly quantitative (e.g., material flow analysis, MFA) which to some extent is useful for assessing the ecological burden of the production of certain goods and activities. Total material requirement is an interesting indicator when it comes to measuring resource productivity as it considers all materials used for a certain product, including indirect material input requirements associated with intermediate imports. A very broad indicator concept—with dozens of sub-indicators—has been developed by researchers at Yale University and Columbia University (Yale 2005) who derive very complex indicators for which equal weights are used. Very complex indicators are, however, rather doubtful in terms of consistency and the message for the general public, industry, and policymakers is often also opaque. Thus one may raise the question of whether a new indicator concept—following the requirements of the OECD (2008) manual and taking into account key economic aspects of green innovation dynamics—can be developed. Before presenting such a new approach, a few general remarks about the SNA are useful to make clear the analytical line of reasoning developed subsequently.

The most common indicator used to assess both economic performance and economic well-being is GDP (GDP: in line with the UN systems of national accounts), which indicates the sum of all newly produced goods and services in a given year. If one wants to consider long-term economic development perspectives, then one would not consider GDP, rather one has to consider NDP (Y') which is GDP minus capital depreciation. Taking into account capital depreciations is important since an economy can maintain its production potential only if the stock of input factors—capital K, labor L, and technology A—are maintained; ultimately, one is only interested in per capita consumption C/L which is the difference of per capita production ($y =: Y/L$) and the sum of private gross investment per capita (I/L) and government consumption per capita (G/L). However, in reality, natural resources R—consisting of renewable and nonrenewables—are also input factors in production. Therefore, "green net domestic product" (GNDP) may be defined here as net national product minus depreciation of natural resources. Indeed to consider such a GNDP is important for many countries which are used to heavily exploiting their respective natural resources. Exploiting nonrenewable resources comes at considerable costs for long-term economic development since running down the stock of nonrenewables implies that future production will decline at some point of time, t.

The World Bank has highlighted the role of the depreciation of natural resources, namely, by calculating genuine savings ratios S'/Y where S' is standard savings S minus the depreciation of capital minus the depreciation of natural resources (and plus expenditures on education as a basis for building the stock of human capital and minus some other elements which are detrimental to sustained economic development—see the subsequent discussion). One should note that there is some positive correlation between GDP per capita and subjective well-being as is shown in recent analysis (Stevenson and Wolfers 2008). Policymakers thus have a strong tendency to emphasize that rising GDP per capita is an important goal. At the same time, it is fairly obvious that the general public is not aware of the difference between GDP and NDP—let alone the significance of NDP and GNDP (sustainable product).

The problem is that although there have been broad international discussions about the greening of national accounts (see, e.g., Bartelmus 2008, 2013), the UN has not adopted any major modernization of its SNA in the past decades. The UN has developed an approach labeled system of integrated economic environmental accounts (SEEA) which, however, has not replaced the standard systems of national accounts. The SEEA basically considers depreciations of natural capital, but the system is rather incomplete as appreciations of natural resources are not taken into account—i.e., the SEEA does not adequately consider improvements in the quality of natural resources (e.g., the water quality of rivers which has improved in many EU countries over time). An interesting indicator to measure the quality of life is the UN human development index which aggregates per capita income, education, and life expectancy. Life expectancy is related to many factors and one may argue that the quality of life is one of them. Another indicator is the index of sustainable economic welfare (ISEW), based on John Cobb (1989), who basically has argued that welfare should be measured on the basis of per capita consumption, value added in the self-service economy (not covered by the SNA) and consumer durables, but that expenditures which are necessary to maintain production should be deducted (e.g., expenditures on health care, expenditures for commuting to work, etc.). The elements contained in the ISEW are not fully convincing, and the policy community has not taken much notice of this.

In the subsequent analysis, it will be argued that one should focus indeed on broader concepts of global sustainability:

- A broader concept should adequately take into account the role of international competitiveness and technological progress—the emphasis on some Schumpeterian elements in sustainability analysis could not only deepen our analytical view of economic–ecological challenges but also help to alert decision-makers in industry and in the policy community to take adequate decisions in the field of innovation and modernization. The focus will be on green international competitiveness as measured by the modified revealed comparative advantage (mRCA), which basically indicates to what extent the respective country has positively specialized on exports—and production—of goods relevant for improving the quality of the environment. Additionally, CO_2 emissions per capita and the role of the genuine savings rate will be considered. Basic aspects of aggregation are taken into account, but no econometric analysis is presented which allows for drawing firm conclusions for the issue of weighing the components of the summary indicator.
- Reconciling economic convergence between the north and south—that is, a declining long-term per capita income gap between the north and the south—and achieving sustained economic growth in the world economy will be easier to achieve if one had a consistent indicator which helps to identify economic–ecological progress and green international leadership.

The basic approach presented here suggests that a new set of indicators is useful for the discussion about global sustainability issues; in this context, energy consumption is one of the key aspects. According to the BP Energy Outlook 2035—published in early 2015—the share of oil in global energy consumption will reduce to a share of about 30 % in 2035 which is 10 points lower than in 2000. The share of coal also will decrease while that of natural gas is expected to increase; the share of renewable energy—including water power—will be close to 15 % in 2035, up from 6 % in 2000. The BP study emphasizes that the energy consumption of OECD countries is likely to grow only very modestly while consumption of newly industrialized countries, including China, will dominate the global growth of energy consumption. From this perspective, energy policy and innovation policy in newly industrialized countries plus north–south energy technology transfer will be crucial for global sustainability. BP expects global greenhouse gas emissions to exceed in 2035 the level of 2013 by about 35 % which would indicate progress of global energy productivity—but it would be insufficient to avoid further global warming. In this context, the new global sustainability EIIW-vita indicator—integrating three key pillars of sustainability (including the share of renewable energy)—gives crucial signals and can contribute to an enhanced green innovation process worldwide. The new metrics proposed is consistent and innovative.

The analysis is organized as follows: Chap. 2 presents standard approaches to environmental degradation and Chap. 3 focuses on green innovation dynamics. Chapters 4 and 5 are on the new indicator concept and the methodology, while Chap. 6 presents results of the EIIW-vita GSI for key countries and Chap. 7 puts the focus on particular aspects in Asia where the majority of the world population is living. To conclude, we draw some critical policy conclusions in Chap. 8.

References

Bartelmus P (2008) Quantitative economics, how sustainable are our economies? Springer, Dordrecht

Bartelmus P (2013) Sustainability economics. Routledge, Abingdon

Bollen J, Guay B, Jamet S, Corfee-Morlot J (2009) Co-benefits of climate change mitigation policies: literature review and new results. OECD Economics Department Working Papers 692

BP (2009) Statistical review of world energy. BP, London

Boyle MA, Kiel KA (2001) A survey of house price hedonic studies of the impact of environmental externalities. J Real Estate Soc 9(2):117–144

Bretschger L (2014), International environmental negotiations. Paper presented at Ausschuss für Außenwirtschaftstheorie und –politik, Augsburg

Cobb CW (1989) The index for sustainable economic welfare. In: Daly J, Cobb CW (eds) For the common good – redirecting the economy towards community, the environment, and a sustainable future. Beacon, Boston

Din A et al (2001) Environmental variables and real estate prices. http://archive-ouverte.unige.ch/unige:5843. Accessed 26 Jan 2015

EPA United States Environmental Protection Agency (2012) A framework for sustainability indicators at EPA. EPA United States Environmental Protection Agency, Washington, DC

Erdem D (2012) Foreign direct investments, energy efficiency, and innovation dynamics. Miner Econ 24(2):119–133

European Commission (2009a) GDP and beyond. Measuring progress in a changing world. COM (2009) 433 final

European Commission (2009b) Sustainable development indicators. overview of relevant FP-funded research and identification of further needs

European Environmental Agency (2008) Energy and environment report 2008. http://www.eea.europa.eu/publications/eea_report_2008_6. Accessed 26 Jan 2015

Francois J et al (2013) Reducing transatlantic barriers to trade and investment. CEPR (for the European Commission), London

Global Footprint Network (2015). www.footprintnetwork.org. Accessed 30 Mar 2015

Hallegatte S, Patmore N, Mestre O, Dumas P, Corfee-Morlot J, Herweijer C, Muir-Wood R (2008) Assessing climate change impacts, sea level rise and storm surge risk in port cities: a case study on Copenhagen. OECD Environment Working Papers 3

Hennicke P, Welfens PJJ (2012) Energiewende nach Fukushima. Deutscher Sonderweg oder weltweites Vorbild?. oekom verlag, München

Hoornweg D, Sugar L, Trejos-Gomez CL (2011) Cities and greenhouse gas emissions: moving forward. Environ Urban 23(2):207–227

ICPR International Commission for the Protection of the Rhine (2014). http://www.iksr.org/. Accessed 26 Jan 2015

Jamet S (2012) Towards green growth in Denmark: improving energy and climate change policies. OECD Economics Department Working Papers 974

OECD (2008) Handbook on constructing composite indicators. OECD, Paris

OECD (2009) OECD territorial reviews: Copenhagen, Denmark. OECD, Paris

OECD (2010) Green growth strategy interim report: implementing our commitment for a sustainable future. Meeting of the OECD Council at Ministerial Level. C/MIN (2010) 5, Paris
OECD (2011) Cities and climate change. OECD, Paris
Pittel K, Lippelt J (2014) Kurz zum Klima: Nachhaltigkeit und Naturkapital – wie viel und wie wichtig? ifo Schnelldienst 1/2014
Sierra Club (2013) The transatlantic free trade agreement: what's at stake for communities and the environment. Sierra Club, Washington, DC
Sprenger RU (1999) Economic globalization, FDI, environment and employment. In: Welfens PJJ (ed) Internationalization of the economy and environmental policy options. Springer, New York
Stern N (2006) The economics of climate change (stern review). HM Treasury, London
Stevenson B, Wolfers J (2008) Economic growth and subjective well-being: reassessing the Easterlin paradox. NBER Paper 14282
Welfens PJJ, Perret JK, Erdem D (2010) Global economic sustainability indicator: analysis and policy options for the Copenhagen process. IEEP 7(2–3):153–185
Welfens PJJ (2010) Transatlantic banking crisis: analysis, rating, policy issues. IEEP 7(3):3–48
Welfens PJJ, Korus A, Irawan T (2014) Transatlantisches Handels- und Investitionsabkommen. Lucius & Lucius, Stuttgart
Welfens PJJ, Irawan T (2014) Transatlantic trade and investment partnership: sectoral and macroeconomic perspectives for Germany, the EU and the US. IEEP 11:293–328. doi:10.1007/s10368-014-0292-9
World Bank (2011) The changing wealth of nations. World Bank, Washington, DC
World Bank (2014) World development indicators database. http://data.worldbank.org. Accessed 21 Jan 2015
Voss E et al (2008) Codes of conduct and international framework agreements: new forms of governance at the company level. http://www.eurofound.europa.eu/publications/ef0792.htm. Accessed 26 Jan 2015
Yale (2005) Environmental sustainability index. Yale Center for Environmental Law and Policy (YCELP) and the Center for International Earth Science Information Network (CIESIN) of Columbia University, Columbia

2 Traditional Approaches to Environmental Degradation and Innovation Theory

Standard approaches to damaging of the environment greatly emphasize the issue of nonrenewable resources. This focus is not surprising, as some vital resources used in industry are important nonrenewable inputs. However, one should not overlook the fact that innovation dynamics and technological progress typically can mitigate some of the problems in the long run—here, the focus is on both process innovations, which economize on the use of resources, as well as product innovations, which might bring about the use of different nonrenewable or of synthetic chemical inputs. At the same time, one may argue that until 2050 there will be considerable global population growth and most of the output growth will come from Asia—including China and India. In these countries, emphasis on fighting global warming is naturally not a top priority, but rather economic catching-up figures prominently in the political system, and economic analysis suggests that China and India still have a large potential for both economic catching-up and long-term growth, respectively (Dimaranan et al. 2009). Nevertheless, one may emphasize that economic globalization also creates new opportunities for international technology transfer and for trade with environmental (green) goods. If there is more trade with green goods and if certain countries successfully specialize in the production and export of such goods, the global abilities in the field of environmental modernization might be sufficient to cope with global warming problems: This means the ability to fight global warming on the one hand and on the other hand the ability to mitigate the effects of global warming. A potential problem of putting more emphasis on innovation dynamics is that a wave of product innovations could trigger additional emissions, which would partly or fully offset the ecological benefits associated with higher energy efficiency which would result in a generally more efficient way to use natural resources.

Sustainability means the ability of future generations to achieve at least the same standard of living that the current generation has achieved. If one adopts a national sustainability perspective, this puts the focus on sustainable economic development in every country of the world economy. Analytical consistency in terms of sustainability imposes certain analytical and logical requirements:

P.J.J. Welfens et al., *Towards Global Sustainability*,
DOI 10.1007/978-3-319-18666-5_2

- As a matter of consistency, one may expect that if there is a group of countries which represents—according to specific sustainability indicators—sustainable development, other countries converging to the same structural parameters of the economy (say per capita income and per capita emissions as well as other relevant parameters) will also be classified as sustainable.
- If all countries are sustainable, then there is sustainability of the overall world economy. What sounds trivial at first is quite a challenge if one considers certain indicators as we shall see.
- An important approach to sustainability has been presented by the World Bank, which calculates genuine savings rates. The basic idea of a broadly defined savings rate is to take into account that current per capita consumption can only be maintained if the overall capital stock—physical capital, human capital, and natural capital—can be maintained. To put it differently: an economy with a negative genuine savings rate is not sustainable. The genuine savings rate concept is quite useful if one is to understand the prospect of the sustainable development of individual countries. The figures presented subsequently basically suggest that OECD countries are well positioned, particularly the USA. This, however, is doubtful, as it is clear that in case the south would converge to consumption patterns of OECD countries—and would achieve economic convergence in terms of per capita income—the world could hardly survive because the amount of emissions and waste would be too large to be absorbed by the earth. For example, the CO_2 emissions would be way above any value considered compatible with sustainability as defined by the IPCC and the Stern report.

The World Bank approach is partly flawed in the sense that it does not truly take into account the analytical challenge of open economies. To make this point clear, let us consider the concept of embedded energy which looks at input–output tables in order to find out what share of the use of energy (and hence CO_2 emissions) is related to exports or net exports of goods and services. For example, the USA has run a large bilateral trade deficit with China—and indeed the rest of the world—for many years and this implies that the "embedded genuine savings rate" (EGSR) of the USA has to be corrected in such a way that the EGSR is lower than indicated by the World Bank. Conversely, China's EGSR is higher than indicated by the World Bank. To put it differently: while the genuine savings rate is indeed useful to assess the sustainability of individual countries at first glance, a second glance, which takes into account the indirect international emissions and indirect running down of foreign stocks of resources (e.g., deforestation in Latin America or Asia due to the net USA/EU imports of goods using forest products as intermediate inputs) related to trade, presents a different perspective; EGSR should not be misinterpreted to take the responsibility away from certain countries; however, EGSR and the genuine savings rate concept—standing for two sides of the same coin—might become a starting point for more green technology cooperation between the USA and China or the EU and China.

Considering the EGSR helps to avoid the misperception that if all countries in the south of the world economy should become like OECD countries, then the overall world economy should be sustainable. According to the World Bank's genuine savings rate, the USA in 2000 was on a rather sustainable economic growth path. However, it is clear that if all non-US countries in the world economy had the same structural parameters—including the same per capita income and the same emissions per capita—as the USA, there would be no global sustainable development. If, however, one considers the EGSRs, the picture looks different. For instance, if one assumes that the EGSR for the USA is lower by 1/5th than the genuine savings rate, it is clear that the USA's position is not as favorable as the World Bank data would suggest.

The ideal way to correct the World Bank genuine savings rate data is to consider input–output and trade data for the world economy so that one can calculate the EGSR; however, such data are available for only a few countries, but in a pragmatic way, one may attribute China's depreciation of natural resources and the CO_2 emissions to the USA and the EU countries as well as other countries vis-à-vis China running a sustained bilateral trade balance surplus. A pragmatic correction thus could rely on considering the bilateral export surplus of China—as an example, if the ratio of total exports to GDP in China is 40 % and if a half of China's export surplus is associated with the USA, then 20 % of China's CO_2 emissions can effectively be attributed to the USA. One might argue that considering such corrected, virtual CO_2 emissions is not really adequate since global warming problems depend on global emissions of CO_2, while individual country positions are of minor relevance. However, from a policy perspective, it is quite important to have a clear understanding of which countries are effectively responsible for what share of CO_2 emissions in the world economy. As sources of CO_2 emissions are both local and national, it is indeed important to not only consider the EGSR but also to know which countries are responsible for what amount of CO_2 emissions.

In the literature, one finds partial approaches to the issue of global sustainability. The concept of the ecological footprint (Wackernagel 1994; Wackernagel and Rees 1996)—as suggested by the WWF (see, e.g., Wiedmann and Minx 2007)—is one important element. The ecological footprint summarizes on a per capita basis (in an internationally comparative way) the use of land, fish, water, agricultural land, and the CO_2 footprint in one indicator, so that one can understand how strong the individual's pressure on the capacity of the earth to deliver all required natural services really is. At the same time, one wonders to what extent one may develop new indicator approaches which emphasize aspects of sustainability in a convincing way.

The global footprint indicator calculated by the World Wildlife Fund (see Fig. 1.1) and its international network indicate the quantitative use of resources for production, namely, on a per capita basis (Global Footprint Network). It is thus a rather crude indicator of the pressure on the global biosphere and the atmosphere. However, it has no truly economic dimension related to international competition and competitiveness, respectively. If, say, country I has the same global per capita footprint as country II, while the latter is strongly specialized in the production and export of green goods—which help to improve the quality of the environment and to increase the absorptive capacity of the biosphere of the importing countries, respectively—the global footprint approach does not differentiate between country I and country II.

If the general public, the private sector, and policymakers are to encourage global environmental problem-solving, it would be useful to have a broadly informative indicator which includes green international competitiveness—see the subsequent analysis. One may argue that positive RCAs for certain sectors are economically and ecologically more important than in other sectors; however, we consider the broad picture across all sectors considered as relevant by the OECD (see list in Appendix). Modified RCAs are particularly useful indicators as they are not distorted by current account imbalances—as is the traditional RCA indicator which simply compares the sectoral export–import ratio with the aggregate export–import ratio (Comtrade database of the United Nations and World Development Indicators/WDI are used in the subsequent calculations).

As regards adjustment dynamics, it is clear that a static view of the economy and world ecological system is not adequate; rather a Schumpeterian innovation perspective is required. In a broader context, environmental upgrading and ecological modernization bring crucial advantages:

- A better environment will stimulate working efforts—and reduce absenteeism due to the illness of workers—so that output will increase.
- A better environment raises the longevity of people: life expectancy can grow and thus lifetime consumption could increase.
- With a declining level of emissions (e.g., SO_2 emissions), there will be less damage to the physical infrastructure and to machinery and equipment: the effective capital depreciation rate will reduce, so that in the context of a neoclassical model, the level of the growth path of per capita income will increase.
- A better environment also improves the utility of consumption—if people are more healthy and can therefore enjoy consumption (in a stochastic environment) more than before, there will be an increase in utility. As one may assume that per capita consumption is proportionate to per capita output, a cleaner environment brings a double benefit. A higher per capita output in the steady state on the one hand and on the other hand the degree of (health) risk in consumption will fall, meaning that investment in a better environment will create a double dividend.
- The ability of a society to switch to a higher environmental quality hinges on investment in research and development in general and on green R&D in particular. Adequate innovation in green technological progress thus is crucial.
- From this perspective, it is useful to take a closer look at key challenges of green R&D.

References

Dimaranan B, Ianchovichina E, Martin W (2009) How will growth in China and India affect the world economy. World Econ 145:551–571

Global Footprint Network. The global footprint network. http://www.footprintnetwork.org. Accessed 26 Jan 2015

Wackernagel M (1994) Ecological footprint and appropriated carrying capacity: a tool for planning toward sustainability. Ph.D. Thesis, The University of British Columbia

Wackernagel M, Rees R (1996) Our ecological footprint: reducing human impact on the earth. New Society, Gabriola Island

Wiedmann T, Minx J (2007) ISAUK research report 07–01. A definition of carbon footprint. http://www.isa-research.co.uk/docs/ISA-UK_Report_07-01_carbon_footprint.pdf. Accessed 26 Jan 2015

Green Innovation Dynamics

3

Innovation dynamics are crucial for achieving sustainable development and, from this perspective, innovation systems and green entrepreneurship are important, as is the institutional framework of economic policy. Resource-saving technological progress and progress in the field of energy efficiency are key challenges. The latter refers to the global warming problem. That such a problem exists can hardly be questioned. Between 1880 and 2010, the global mean temperate increased and greenhouse gas emissions have contributed to this (see Fig. 3.1).

From a theoretical perspective, CO_2 emissions can be rewritten in the following decomposition (with *E* denoting energy, *Y* real output, *N* population): $CO_2 = (CO_2/E)(E/Y)(Y/N)N$. The first term is the carbon intensity in the use of energy. The term E/Y is energy intensity, Y/N is gross domestic product per capita, and *N* is the population. We already know that the growth in world population between 2010 and 2050 will be 42.9 %, if one assumes that the rise of *N* will be from seven billion to ten billion. Global per capita income is likely to increase by between 2 and 3 % per year; if it were 3 % over 75 years, we get a roughly ninefold increase in one generation. From this perspective, it is clear that CO_2 emissions can be stabilized only if CO_2 intensity and the energy intensity can be starkly reduced. A reduction in CO_2 intensity can be achieved through an increase of energy efficiency (reducing E/Y) and through a rise of the share of renewable energy. The latter will bring about a decarbonization of energy use and a fall of CO_2/E, respectively. As can be seen from the subsequent graph, there has been continuous growth of the world income. A rise of per capita income has several aspects in the context of sustainability:

- The use of energy and hence the level of CO_2 emissions might increase. At the same time, a rising per capita income is typically associated with a higher stock of knowledge and of physical capital and this should facilitate the solving of environmental (and other) problems. Clearly, a sustainable economy will require that there is sufficient reinvestment in the capital stock; otherwise current generations would undermine the consumption opportunities of future generations. Thus sufficient savings will be necessary, as savings are the basis

P.J.J. Welfens et al., *Towards Global Sustainability*,
DOI 10.1007/978-3-319-18666-5_3

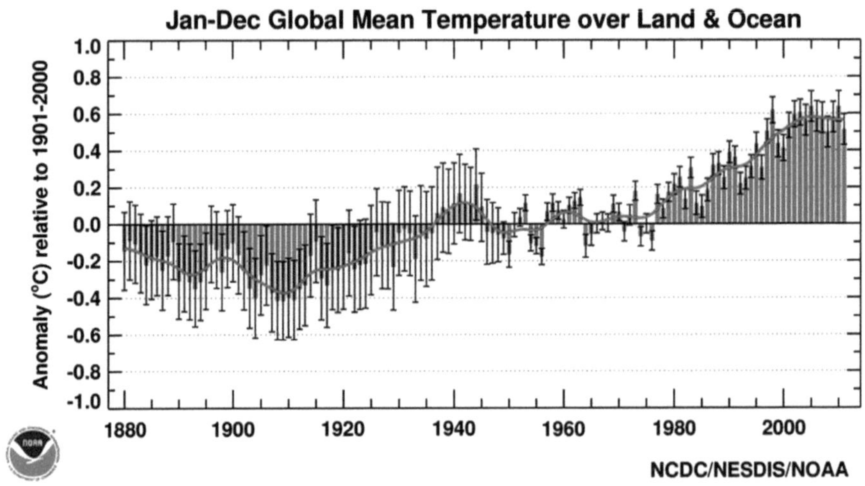

Fig. 3.1 Mean temperature over land and ocean. *Data Source*: National Climatic Data Center

for financing gross investment: the sum of reinvestment and net investment (increase of the capital stock K).

- The demand for differentiated products will increase and such products are typically associated with product innovations; most innovations in the world occur in the tradable goods sector and hence the share of green products in exports is quite important from a sustainability perspective. From a theoretical perspective, one may argue that the higher the ratio (RCA) of green exports relative to green imports—relative to the overall export–import ratio—in country I is, the higher the international specialization in green products is and indeed the more the country contributes to solving the environmental problems of trading partners. If the relative export position in environmentally friendly products is strong, the range of products sold in the home market also should be rather environmentally friendly.

Previously, it was emphasized in Welfens et al. (2010) pp. 12–13 that:

Aghion et al. (2009) argue that radical innovations are needed to bring about strong progress in CO_2 emissions: Given the fact that the share of green patents in total global patents is only about 2 %, one cannot expect that incremental changes in technologies will bring about strong improvements in energy efficiency and massive reductions of CO_2 emissions per capita; while the generation of electricity is a major cause of CO_2 emissions the share of R&D expenditures in the sector's revenues was only 0.5 %.

The Kyoto and Bali conferences on climate policy have forcefully asserted that the industrialized countries should support developing countries in the field of technological progress, and flexible instruments, such as clean development mechanisms, joint implementation and emission trading implemented by the Kyoto Protocol, are already contributing to the efficient use of resources in the global economy. However, sharp national productivity differences will remain, explaining a large part of the difference in national incomes. At this point, technology plays an important role in shaping productivity.

With connection to this, technology and know-how infusions from abroad are indispensible for sustainable growth in developing countries. Keller (2004) emphasizes that FDI and international trade are crucial channels of technology diffusion. The evidence is easy to see in terms of imported goods and services, which stand for embodied progress and new knowledge, respectively. Nevertheless, it is fairly obvious that domestic investment in R&D and technology is also necessary. While OECD countries might be willing to transfer certain technologies to the South, it also is necessary that the adaptation of know-how be strongly improved in the South: Developing countries can only benefit from a technology transfer if they reach a minimum level of human capital, which, again, requires investment in education (Xu 2000). Kemfert (2002) stresses that integration of technological change in a multi-regional trade system improves energy efficiency and could reduce environmental problems. In this context, flexible instruments facilitate technological progress and technology transfer, respectively: This, in turn, increases the prosperity in the host countries, where positive knowledge and spillover effects play an important role (in particular in the developing countries, they also lead to improved international competitiveness—at least if strong inflows of foreign direct investment can be achieved). In this context, the channels for improvements in resource use and greenhouse mitigation through technology transfer can be considered to be a broad range of relevant aspects: technology transfer per trade of goods and services, FDI, international programs and development aid (Peterson 2008). In an increasingly internationalized supply side setting, domestic firms can realize improvements in energy efficiency and decrease their energy intensity by investing in new technologies. Higher productivity results from the spillover of advanced technologies and educational improvement, but also from advanced management skills. A number of important questions arise from these reflections: Can free international trade help to increase efficient use of resources? The WTO (1999) has published only one study on trade and environmental problems—which is rather disappointing for such a large international organization.

A positive answer to the question requires a rising technology level, successful restructuring of production processes and a higher level of competition. In this context, the relationship between the internationalization of economies and environmental sustainability has been a key issue since the late 1970s, and interest in the topic has increased tremendously since the 1990s; particularly in the wake of the argument of Grossman and Krueger (1995), assuming that globalization causes economic growth, the relationship between globalization and environmental quality is not negative. On the contrary, positive effects of economic growth on the environment can be observed for most environmental quality indicators. In addition, a national income per capita of $8000 a turning point for increasing environmental quality. Similar findings in the field of sulphur dioxide pollution problems have been provided by Antweiler et al. (2001): technology transfer is coupled with the effects of scale created by international trade reduction e.g. sulphur dioxide pollution.

Green innovation dynamics can be analyzed in several dimensions:

- Input dimension (research and development: R&D): Here, one will focus on green R&D projects and R&D expenditures related to developing new environmentally friendly products, respectively.
- Output dimension: The number of patents and of scientific publications in the relevant field.
- Indirect impact measures can be derived from aggregate data—based on a standard model: e.g., improvements in resource efficiency and resource productivity using standard decomposition analysis.

If one looks at various dimensions of sustainability, one should not follow a popularly held view that only industrialized countries can be leaders in key fields of sustainability. The following table shows that the top 20 countries for the share of renewable electricity are mostly made up of developing countries and some newly industrializing countries plus Norway. Several OECD countries are certainly fairly well positioned. However, in a global context, it is, of course, very important how the big countries, i.e., the USA, China, Russia, India, the EU (a quasi country), and Indonesia plus Japan, are positioned. The share of renewables in electricity in China was close to 20 % in 2014, 4/5th of which is derived from hydroelectrical power generation. The small share of about 4 % for other renewables points to an enormous potential for China to increase its share of renewables over the coming decades.

As regards the top 20 countries in the production of electricity from renewable sources in 2011, seven of the top 10 countries were developing countries. It is remarkable that only two OECD countries were among the top 20 countries (see Table 3.1).

As regards environmentally friendly products, there is a classification by the OECD which is shown in the appendix. To the extent that the innovations of firms are aimed at reducing resource intensity, raising the longevity of the product or reducing energy efficiency, the respective new products may be classified as

Table 3.1 Top 20 countries in the production of electricity from renewable sources in 2011

	Top 20 countries	Renewable share in electricity production (%)
1	Paraguay	100.00
2	Iceland	99.99
3	Albania	99.98
4	Nepal	99.91
5	Mozambique	99.88
6	Zambia	99.65
7	Ethiopia	99.36
8	Tajikistan	98.82
9	Namibia	98.18
10	Norway	96.53
11	Kyrgyz Republic	93.28
12	Costa Rica	91.22
13	Brazil	87.12
14	Colombia	82.36
15	Georgia	77.40
16	New Zealand	75.85
17	Togo	75.54
18	Sudan	75.19
19	Zimbabwe	74.40
20	Cameroon	74.37

Data Source: World Bank, World Development Indicators

environmentally friendly. The definitions chosen by the OECD naturally leave some room for criticism.

Green innovativeness is part of the overall Schumpeterian dynamics of a modern economy. In the literature, firms' strategic positioning and specific managerial factors contribute to raising innovativeness (Borghesi et al. 2012; De Marchi 2012; Horbach 2008; Mazzanti and Zoboli 2012). It is important to have a consistent definition of eco-innovations (e.g., Arundel and Kemp 2009; Kemp and Pearson 2007; OECD 2009). It is important to make a distinction between policy-induced eco-innovations (Hottenrott and Rexhäuser 2013; Brunnermeier and Cohen 2003; Cleff and Rennings 1999; Porter and Van Der Linde 1995) and price-induced eco-innovations (Popp 2002; Newell et al. 1999; Jaffe and Palmer 1997; Lanjouw and Mody 1996).

There are two basic definitions of eco-innovations: OECD (2009) and Kemp and Pearson (2007):

- OECD (2009): "The creation of new, or significantly improved, products (goods and services), processes, marketing methods, organizational structures and institutional arrangements which lead to environmental improvements compared to relevant alternatives."
- Kemp and Pearson (2007): "The production, assimilation or exploitation of a product, production process, service or management or business method that is novel to the organization (developing or adopting it) and which results, throughout its life cycle, in a reduction of environmental risk, pollution and other negative impacts of resources use (including energy use) compared to relevant alternatives."

It is interesting to consider eco-innovation-related patterns which show the leading position of Denmark, Germany, Sweden, Finland, and the Netherlands, if one looks at patterns per capita in 2008. An alternative perspective is to focus on eco-innovation-related publications. The scientific publications per capita in 2011 were highest in Finland, Luxembourg, and Sweden. These three countries also had a leading position in 2010—with Denmark being a close follower (see Fig. 3.2, 3.3, 3.4, and 3.5).

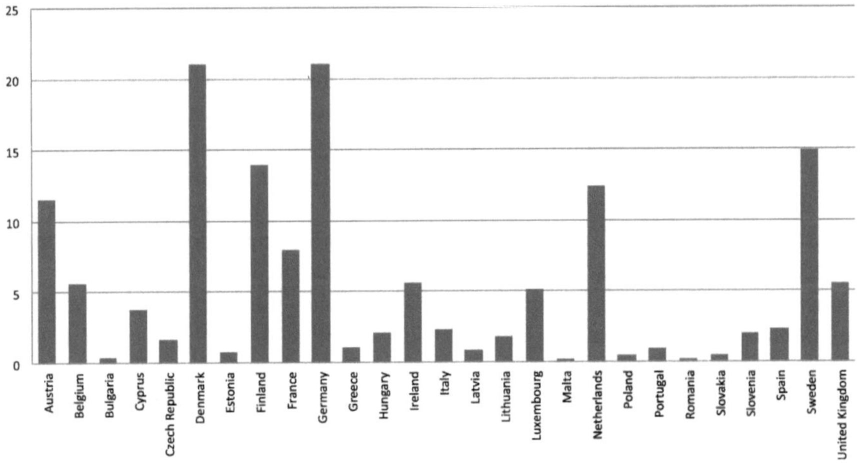

Fig. 3.2 Eco-innovation-related patents (per million population) in 2008. *Data Source*: Eco-Innovation Observatory (2014)

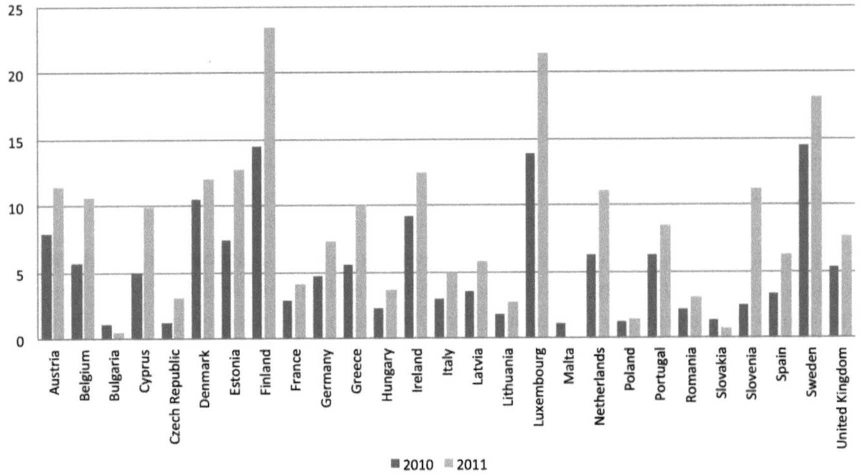

Fig. 3.3 Eco-innovation-related publications (per million population). *Data Source*: Eco-Innovation Observatory (2014)

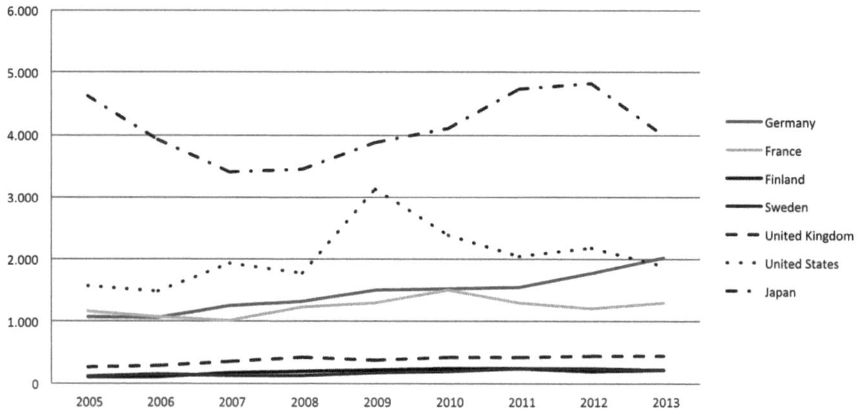

Fig. 3.4 Government environmental and energy R&D (million euros). *Data Source*: Eurostat

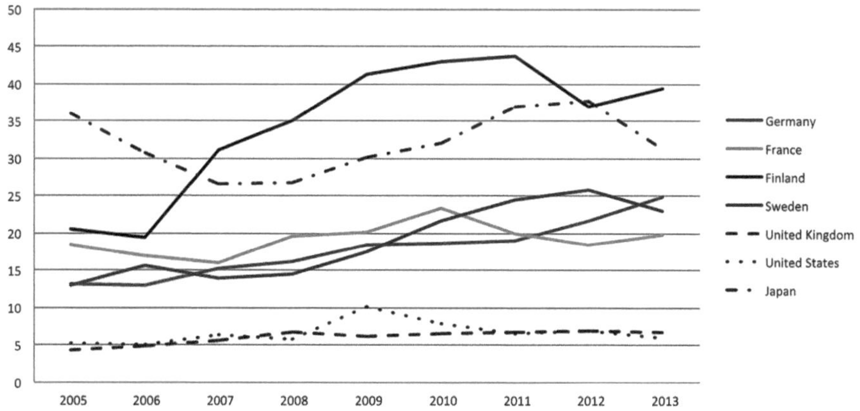

Fig. 3.5 Government environmental and energy R&D (euro per inhabitant). *Data Source*: Eurostat

References

Aghion P et al (2009) No green growth without innovation. Bruegel Policy Brief, Brussels

Antweiler W, Copeland BR, Taylor MS (2001) Is free trade good for the environment? Am Econ Rev 91(4):877–908

Arundel A, Kemp R (2009) Measuring eco-innovation. UNU-MERIT Working Papers 2009-017

Borghesi S, Cainelli G, Mazzanti M (2012) Brown sunsets and green dawns in the industrial sector: environmental innovations, firm behavior and the European emission trading. Working Papers 2012 03

Brunnermeier SB, Cohen MA (2003) Determinants of environmental innovation in US manufacturing industries. J Environ Econ Manag 45(2):278–293

Cleff T, Rennings K (1999) Empirische Evidenz zu Besonderheiten und Determinanten von Umweltinnovationen. In: Rennings K (ed) Innovation durch Umweltpolitik, vol 36, - ZEW-Wirtschaftsanalysen. Baden-Baden, Nomos, pp 47–99
De Marchi V (2012) Environmental innovation and R&D cooperation: empirical evidence from Spanish manufacturing firms. Res Policy 41:614–623
Eco-Innovation Observatory (2014). http://www.eco-innovation.eu. Accessed 26 Jan 2015
Eurostat. http://ec.europa.eu/eurostat. Accessed 26 Jan 2015
Grossman GM, Krueger A (1995) Economic growth and the environment. Q J Econ 110 (2):353–377
Horbach J (2008) Determinants of environmental innovation-new evidence from German panel data sources. Res Policy 37:163–173
Hottenrott H, Rexhäuser S (2013) Policy-induced environmental technology and incentive efforts: is there a crowding out? ZEW Discussion Papers 13-115
Jaffe AB, Palmer K (1997) environmental regulation and innovation: a panel data study. NBER Working Paper 5545
Keller W (2004) International technology diffusion. J Econ Lit 42(3):752–782
Kemfert C (2002) Induced technological change in a multi-regional, multi-sectoral integrated assessment model (WIAGEM); impact assessment of climate policies. Department of Economics, University of Oldenburg, Oldenburg
Kemp R, Pearson P (2007) Final report MEI project about measuring eco-innovation, UNI-Merit
Lanjouw JO, Mody A (1996) Innovation and the international diffusion of environmentally responsive technology. Res Policy 25:549–571
Mazzanti M, Zoboli R (2012) A political economy approach to resource taxation: weak sustainability, revenue recycling and regional planning. Working Papers 2012 02
Newell RG, Jaffe AB, Stavins RN (1999) The induced innovation hypothesis and energy-saving technological change. Q J Econ 114(3):941–975
OECD (2009) Sustainable manufacturing and eco-innovation: towards a green economy. OECD, Paris
Peterson S (2008) Greenhouse gas mitigation in developing countries through technology transfer? – a survey of empirical evidence. Mitig Adapt Strateg Glob Chang 13(3):283–305
Popp D (2002) Induced innovation and energy prices. Am Econ Rev 92(1):160–180
Porter ME, Van Der Linde C (1995) Toward a new conception of the environment-competitiveness relationship. J Econ Perspect 9(4):97–118
Welfens PJJ, Perret JK, Erdem D (2010) Global economic sustainability indicator: analysis and policy options for the Copenhagen process. EIIW Discussion Paper 174
WTO (1999) Trade and environment. WTO, Geneva
Xu B (2000) Multinational enterprises, technology diffusion, and host country productivity growth. J Dev Econ 62(2):477–493

New Indicator Concept

4

At a basic level, one could build indicators based on the individual, which is often a good way to motivate individuals to reconsider their respective lifestyles. Alternatively (or in a complementary way), one may develop indicators with a focus on individual countries so that the focus is more on political action, including opportunities for international cooperation. A consistent theoretical basis for a global sustainability indicator is useful, and it is therefore argued here that one should focus on three elements for assessing global sustainability. An indicator set will be suggested in which the main aspects are:

- Ability to maintain the current standard of living based on the current capital stock (broadly defined). Hence, "genuine savings rates"—including the use of forests and nonrenewable energy sources—are an important aspect. To the extent that countries are unable to maintain the broader capital stock (including natural resources), there is no sustainable consumption to be expected in the long run.
- Ability to solve environmental problems: If we had an adequate subindicator related to innovation dynamics, the composite sustainability indicator would then have a true economic forward-looking dimension. If countries enjoy a positive revealed comparative advantage in the export of environmental products ("green goods"), one may argue that the respective country contributes to the global solving of environmental problems. As it has specialized successfully in exporting environmental products, it is contributing to improving the global environmental quality; also, countries which have specialized in exports of green goods may be expected to use green goods intensively themselves—not least because of the natural knowledge advantage in producer countries and because of the standard home bias of consumers. Countries will be ranked high if they have a high-modified RCA (mRCA) in green goods: The mRCA for sector i is defined in such a way that the indicator is zero if the respective sector's export share is the same as that of all competitors in the world market and it is

P.J.J. Welfens et al., *Towards Global Sustainability*,
DOI 10.1007/978-3-319-18666-5_4

normalized in a way that it falls in the range −1, 1 (with positive values indicating an international competitive advantage).

- Pressure on the climate in the sense of global warming. Here, CO_2 emissions are clearly a crucial element to consider. The share of renewables could be an additional element, and a rising share over time would indicate not only an improvement of the environmental quality—read less pressure re global warming—but also reflect "green innovation dynamics." In principle, the aggregation of subindicators should use a weighing scheme based on empirical analysis.
- A composite indicator can conveniently summarize the various dimensions to be considered, and indeed this is done subsequently.

For a group of countries, the genuine savings rate and the gross domestic savings rate are shown for the year 2000 which is a basic year of reference. The definition of net national savings is gross national savings minus capital depreciation (consumption of fixed capital); if we additionally add education expenditures and subtract energy depletion, mineral depletion, net forest depletion, PM10 damage (particulate matter), and CO_2-related damage, one gets the genuine savings rate. From a sustainability perspective, the savings rate indicates a high degree of sustainability if it is high enough to finance the reinvestment in machinery and equipment—plus infrastructure—plus hypothetical reinvestment in the stock of natural (renewable) resources; plus human capital formation.

Thus, sustainability is weak—based on standard World Bank data—if the genuine savings rate is relatively low (see Fig. 4.1). This is particularly the case for Azerbaijan, Kazakhstan, Iran, Saudi Arabia, and Russia. The latter two are in a

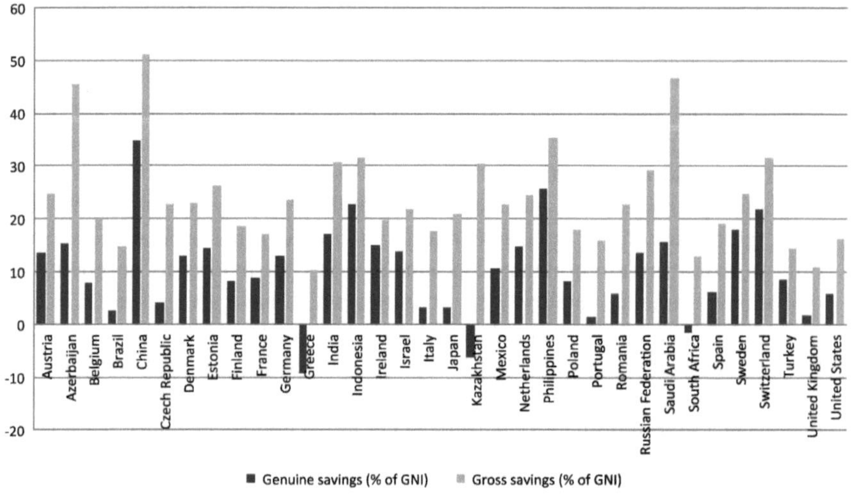

Fig. 4.1 Genuine savings versus gross savings (% of GNI), 2012 *Data Source*: World Bank (2014), World Development Indicators

very weak position, as the respective genuine saving rates are negative, having exceeded −10 %. Moreover, it is noteworthy that for many countries, there is a large gap between the standard savings rate and the genuine savings rate. This suggests that with respect to economic sustainability, there is a veil of ignorance in the broader public and possibly also among policymakers.

A crucial dimension of global sustainability is CO_2 emissions per capita; this indicator is mainly related to the use of energy for production and consumption, respectively. The share of renewables is also a crucial element for climate policies. The energy sector, however, is subject to considerable relative price shifts over time and indeed has reacted with too strong price shocks with innovations. High and rising oil prices have undermined global economic dynamics in the period from 2006 to 2008, and representatives of industry and OECD countries have raised the issue as to how, why, and how long such price increases will continue. While it seems obvious that sustained relative price changes should stimulate innovation—see the analysis of Grupp (1999) for the case of the OPEC price shocks of the 1970s—as well as substitution effects on the demand side and the supply side, it is rather unclear which mechanisms shape the price dynamics in the short term and the long run. The following analysis takes a closer look at the issues, presents new approaches for economic modeling, and also suggests new policy conclusions.

In the wake of the two oil price shocks of the 1970s—each bringing with it a quadrupling of the oil price—the economics of exhaustible resources became an important research field (e.g., Stiglitz 1974; Dasgupta and Heal 1979; Sinn 1981). Oil and gas are particular examples of nonrenewable resources, and they are politically sensitive as the main deposits are concentrated regionally, in the case of oil in politically rather sensitive Arab countries as well as Iran and Russia. In addition, major oil producers have established OPEC, which became a powerful cartel in the 1970s when it controlled about 60 % of the world market for oil. As transportation costs for oil are very small, the oil price is a true world market price since equilibrium is determined by world oil supply and global oil demand. There is considerable short-term oil price volatility in the short run, and there have been major shifts in oil prices over the medium term. Changes in market structure will affect the optimum rate of depletion of resources (Khalatbari 1977).

The oil and gas sector has a long history of high Schumpeterian dynamics. Analysis by Enos (1962) suggests that there is a time lag of about 11 years between invention and innovation. By implication, R&D promotion in this industry will go along with considerable time lags with respect to innovation—this is also a challenge for policymakers, who would have to apply a relatively long time horizon. As regards R&D promotion, Furtado (1997) found that differences in the degree of appropriability between upstream and downstream of the oil industry had a great impact on the effect of R&D promotion. There are regional case studies on the dynamics of innovation in the oil and gas industry—concerning Stavanger and Aberdeen (Hatakenaka et al. 2006)—which show that different approaches to R&D promotion can have similar effects. It is also noteworthy that the energy sector has been a leading early user of information technology (Walker 1986).

A rising relative price of nonrenewables is often considered inevitable, due to long-term global population growth and also high aggregate output growth since the 1990s in the world economy. The use of fossil energy sources does not only have economic issues at stake, but it is also relevant in terms of global warming issues. The Stern Report (Stern et al. 2006; Nordhaus 2006; Latif 2009) has raised international attention about the dynamics of the use of energy and the associated CO_2 emissions, as have the policy activities and UN reports with a focus on the Kyoto Protocol. There is long-term concern that high economic global growth will strongly stimulate the demand for energy and hence raise emissions. At the same time, there are also medium-term concerns about the potential negative impact of oil price shocks. While higher real oil prices might be useful at encouraging a more efficient use of energy resources, there could also be inflation and unemployment problems linked to sudden rises of nominal oil prices.

As regards CO_2 emissions per capita, we see a well-known picture in which the United States was leading with a relatively poor performance up to 2000 (see Fig. 4.2).

As regards the consistent composite indicator (with adequate centering), a positive position is strictly defined as a favorable global position, and a negative value reflects ecological weakness and to some extent a lack of green innovativeness or inefficiencies in the use of energy-intensive products (as mirrored in the CO_2 per capita indicator); more and better innovations can improve the position of the composite indicator so that the main message is that green innovation dynamics matter—thus government should encourage green Schumpeterian dynamics, particularly if there are positive national or international external effects. Specialization in green knowledge-intensive industries and

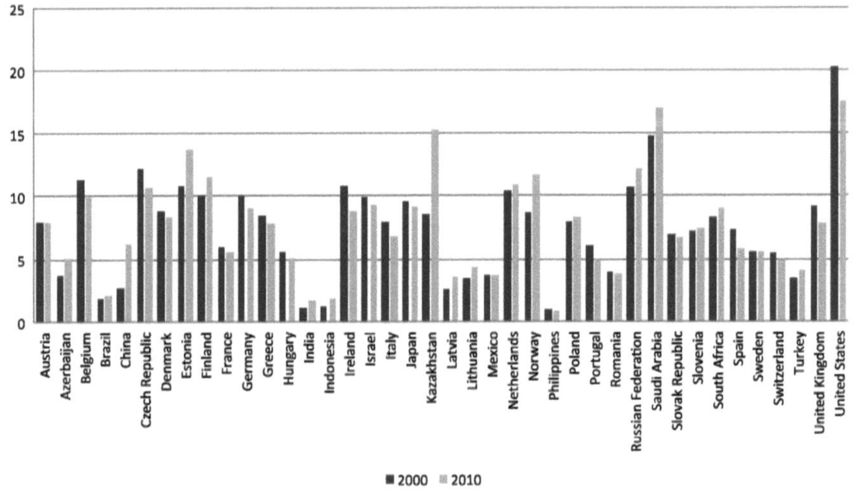

Fig. 4.2 CO_2 emissions (metric tons per capita) *Data Source*: World Bank (2014), World Development Indicators

positive green RCAs could go along with national or international positive external effects; however, there are few empirical analyses available here. The aggregate indicator shows results which, of course, are somewhat different from the simple aggregation procedure; one can clearly see that careful standardization is required for consistent results.

From a methodological point, the weights attached to the individual components of the indicator could be determined through empirical analysis. Factor loadings are useful starting points for a valid approach. It should be emphasized that the new indicator set proposed (even disregarding the weighing issue) puts the analytical and policy focus on the issue of global sustainability in a new way. The indicator emphasizes long-term opportunities and global sustainability. While this approach is only a modest contribution to the broader discussion about globalization and sustainability, it nevertheless represents analytical progress. There is little doubt that specific issues of sustainability—for example, global warming (see Appendix)—will attract particular interest from both the media and political systems. At the same time, one may emphasize that the new broad indicators developed are useful complements to existing sustainability indicators such as the global footprint from the WWF.

The indicator presented is complementary to existing sustainability indicators. However, it has two specific advantages:

- It emphasizes within the composite indicator a dynamic view, namely, the Schumpeterian perspective on environmental product innovations.
- It is in line with the OECD handbook on composite indicators.

The indicator for SO_2 emissions can be easily aggregated for global emissions, while the genuine savings indicator cannot easily be aggregated if one wants to get global sustainability information. However, as regards the genuine savings indicator, one may argue that if the population-weighted global savings indicator falls below a critical level, there is no global sustainability. One might argue that the global genuine savings rate—a concept which obviously does not need to focus on embedded (indirect) use of materials and energy—should reach at least 5 %, otherwise there is a risk that adverse economic or ecological shocks could lead to a global genuine savings rate which is close to zero; and such a situation could, in turn, lead to economic and political international or national conflicts which could further reduce genuine savings rates in many countries so that global sustainability seems to be impaired.

There are many further issues and aspects of the indicator discussion which can be explored in the future. One may want to include more subindicators and also to consider robustness tests, namely, whether changing the weights of individual subindicators noticeably changes the ranking of countries in the composite index.

Strategic Views

Global warming represents a long-term problem which is related mainly to the use of fossil-based energy resources. The Kyoto protocol established an international framework which—excluding the USA and Australia (the latter signed the Protocol in 2008) as the only OECD countries—imposed restrictions on industrialized countries aimed at reducing greenhouse gases by a certain percentage by 2012. EU countries have adopted an emissions trading system which establishes a certain price for CO_2 emission permits. Energy producers and energy-intensive producers will have to buy such permits unless they obtained them in the first allocation period. Firms will adjust production in such a way that the marginal costs of avoiding CO_2 emissions equal the market price of the emission permit. With a uniform price of emission permits, the marginal costs of CO_2 reduction will be equal across firms. Climate policy measures undertaken by firms or government always have opportunity costs, and the EU's approach of introducing an emission trading system (ETS) is an efficient way at achieving the politically desired reduction of CO_2 emissions; the EU has set a certain overall cap for the EU itself, and the individual countries have made commitments concerning the reduction of national CO_2 emissions. Analyses by economists (e.g., Ward 2006; Pearce 1999; Weimann 1995) have emphasized that a Pigou tax or an international ETS may both be considered as equivalent instruments.

From a theoretical perspective, an ETS seems better than a Pigou tax, namely, since ETS is effectively like a flexible international Pigou tax. One may, however, raise some doubts about the effectiveness of ETS, namely, to the extent that the price of emission certificates is linked to general asset market dynamics—the international banking crisis of 2007/2008, which created some (downward) overshooting, has caused emission permit prices to fall drastically in 2008/2009 and in the 5 years after the collapse of Lehman Brothers.

Germany has adopted a specific law (the Law on Renewable Energy, *Erneuerbare Energie Gesetz*) which subsidizes wind energy generation and solar energy generation, both considered useful ways to reduce CO_2 emissions. This at least was emphasized by several German governments, which also pointed out that many new jobs had been created by the expansion of the solar panel industry and the production of wind mills and related software.

Taking a closer look at the economic aspects of the subsidization of renewables, a modified perspective is adequate. Indeed, a rather inefficient way of reducing CO_2 emissions concerns solar energy generation. In the first two trading periods, which started in 2005 and 2008, German firms obtained emission permits for free. While the price of emission permits in the EU has hovered around 20 € per ton, the costs of avoiding 1 ton of CO_2 emissions through solar energy production (in Germany households producing solar power are guaranteed a price of about 50 cent per kWh, while the market price is only 20 cent per kWh) is in the range of 700–1200 €, and wind power generation amounts to costs of 100–200 € per ton of CO_2 avoided (Weimann 2009). To the extent that the German government subsidizes solar power and wind power and CO_2 emissions in Germany are reduced, there is no improvement in global CO_2 emissions. The German improvement simply leads to a surplus of CO_2 emissions in Germany, so that the German demand/EU demand for emissions permits reduces; the price of emission certificates will fall and energy-intensive producers in other EU countries will thus increase their emissions.

There are, however, some arguments for a modified and low subsidization of wind energy generation and solar energy generation. Two specific elements are required:

- Subsidies should only refer to projects based on new technologies; thus, innovations in energy-related fields would be subsidized—and this is useful to the extent that there positive external effects.
- Static and dynamic scale economies could also be an argument for subsidizing wind energy and solar energy. To exploit such economies of scale in a world economy in which product innovations fetch a Schumpeterian premium price can be a useful way of obtaining rents in international markets. The more people switch to CO_2-free energy sources—which partly might reflect prestige effects on the demand side—in an ever growing number of countries, the higher the potential rents which can be appropriated through first-mover advantages.

From an empirical perspective, one should like to know how important static and dynamic scale economies as well as positive external effects of innovations are. Since the global warming problem refers to CO_2 emissions and other greenhouse gases in a worldwide perspective, it is not efficient to reduce emissions of greenhouse gases in particular countries through particular national subsidies. A global approach to establishing an ETS would be useful. However, one may emphasize that the stabilization of financial markets should be achieved first, as otherwise a very high volatility of certificate prices is to be expected; future markets for such certificates should also be carefully developed, and it is not obvious that such markets will necessarily be in the USA; the EU has a certain advantage here, as the EU has taken a lead in the trading of emission certificates. There are policy pitfalls which one should avoid in setting up ETS; for example, the German government largely exempted the most energy-intensive sectors in the first allocation period—those sectors would normally have rather big opportunities to achieve cuts in energy intensity and CO_2 emissions, respectively; Klepper and Peterson (2006) have calculated that the welfare loss of emission trading could have been 0.7 % of GDP in the first German National Allocation Plan, while in reality the welfare gain amounted to about 2.5 % of GDP.

Government incentives on renewables could be a useful element of environmental modernization. However, the feed-in tariff approach realized in Germany—with a feed-in tariff for solar energy and wind energy that is fixed over 20 years—is rather doubtful, as there are poor incentive effects for green innovations. The implicit subsidy for investors in solar power and wind energy is financed through a special surcharge on household electricity prices and a surcharge for the majority of small and medium firms; large electricity consumers in industry, particularly in the tradables sector, are exempt from the surcharge. Specific incentives for

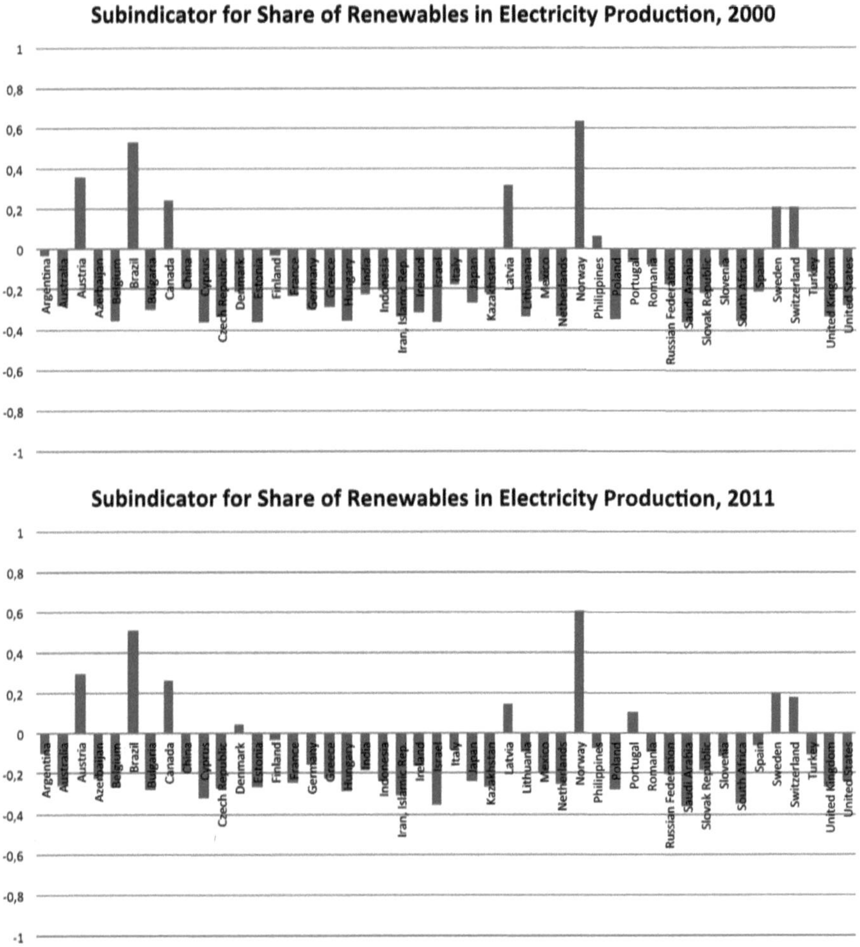

Fig. 4.3 Normalized indicator on the share of renewables in selected countries: 2000 versus 2011. Benchmark = World average *Data Source*: World Bank (2014), World Development Indicators, EIIW calculations

innovation could be expected only if the feed-in tariff would be tied to electricity market prices: as prices fall—for example, due to a rising supply of renewable energy—the implicit subsidization of solar power and wind power (plus bioenergy) should be reduced.

As regards the share of renewables used in energy generation, the following tables show that there are large differences across various countries. Following the general approach presented here—with the world average set at zero (and the indicator normalized in a way that it falls in the range 0, 1)—we can see that there are some countries which are positively specialized in renewable energy: Austria, Brazil, Finland, India, Italy, Latvia, the Philippines, Portugal, Sweden,

Switzerland, and Turkey, all have positive indicators. It is noteworthy that the position of Azerbaijan, Iran, Kazakhstan, the Netherlands, Russia, and the UK is clearly negative. Comparing 2000 and 2011, the negative position of China is remarkable, while at the same time, the UK, slightly, and Germany, strongly, have improved their respective positions (see Fig. 4.3). There is no doubt that countries such as Russia and China could perform much better in the field of renewables provided that government encourages innovative firms and innovations in the renewable sector on a broader scale.

References

Dasgupta P, Heal G (1979) Economic theory and exhaustible resources. Cambridge University Press, Cambridge

Enos JL (1962) Invention and innovation in the petroleum refining industry. In: The rate and direction of inventive activity: economic and social factors, universities-national bureau committee for economic research and the committee on economic growth of the social science research councils. Princeton University Press, Princeton, NJ, pp 299–321

Furtado A (1997) The French system of innovation in the oil industry: some lessons about the role of public policies and sectoral patterns of technological change in innovation networking. Res Policy 25(8):1243–1259

Grupp H (1999) Environment-friendly innovation by price signals or regulation? An empirical investigation for Germany. Jahrb Natl Stat 219(5–6):611–631

Hatakenaka S, Westnes P, Gjelsvik M, Lester RK (2006) The regional dynamics of innovation: a comparative study of oil and gas industry development in Stavanger and Aberdeen. Paper presented at the SPRU 40th anniversary conference, University of Sussex

Khalatbari F (1977) Market imperfections and the optimum rate of depletion of natural resources. Economica 44:409–414

Klepper G, Peterson S (2006) Emission trading, CDM, JI, and more – the climate strategy of the EU. Energy J 27:1–26

Latif M (2009) Klimawandel: Hintergründe und Zukunftsszenarien. Paper presented at the conference climate change – status and perspectives of Verein für Sozialpolitik, Magdeburg, 9 Sept 2009

Nordhaus WD (2006) The stern review on the economics of climate change. NBER Working Paper W12741

Pearce E (1999) Economics and environment. Edgar Elgar, Northampton

Sinn H-W (1981) Stock dependent extraction costs and the technological efficiency of resource depletion. Z Wirtsch Sozialwissen 101:507–517

Stern N et al (2006) The economics of climate change (stern review). HM Treasury, London

Stiglitz JE (1974) Growth with exhaustible natural resources: efficient and optimal growth paths. Rev Econ Stud 41:123–137 (Symposium on the economics of exhaustible resources)

Walker W (1986) Information technology and energy supply. Energy Policy 23:466–488

Ward FA (2006) Environmental and natural resource economics. Pearson Prentice Hall, Upper Saddle River

Weimann J (1995) Umweltökonomik: Eine theorieorientierte Einführung, 3rd edn. Springer, Berlin

Weimann J (2009) Klimapolitik im Zeitalter des Emissionshandel. Wirtschaftswissenschaftliches Studium 38(2):86–90

World Bank (2014) World development indicators database. http://data.worldbank.org. Accessed 21 Jan 2015

5 The Methodology of the EIIW-vita Indicator

There are many sustainability approaches applied in the literature and naturally one will want to carefully consider the methodology being applied. Ultimately, a composite sustainability indicator which takes several dimensions of sustainability into account should, at the very least, be consistent, informative, and relevant, for households, investors, and policymakers alike. Ideally, one will want to be able to aggregate national indicators across a number of countries; thus, one could also have access to information on global sustainability. It will be important to choose those elements of a composite indicator which give reliable information about the sustainability of production on the basis of current economic patterns, while also taking into account the international dimension of production and consumption. An examination of the methodology of indicator building is a necessary part of a comprehensive analysis.

The preceding chapters have already shown that sustainability is a multidimensional concept, which cannot be condensed into a simple, one-line definition. Accordingly, an indicator trying to capture the essential aspects of sustainability needs to encompass a number of select variables which cover more than one of the characteristics of sustainability.

On the other hand, the question can be asked of how detailed a suitable measure for sustainability has to be. The main concern in this chapter lies in creating a composite indicator which is simple enough, so that each subindicator used in its creation adds significantly to its explanatory power, while still needing to include enough partial indicators to cover a number of important aspects of sustainability and thereby assure the high validity required of a composite indicator.

Aside from this central goal, the indicator needs to account for the following aspects that are based in part on EPA (2012) and SEDAC (2007):

- Theoretical and empirical foundation

  The indicator and all of its subindicators need to be founded on sustainability, by way of theoretical or empirical insights. Using established indicators assures that the indicator is not generated by chance, based on data which has been

P.J.J. Welfens et al., *Towards Global Sustainability*,
DOI 10.1007/978-3-319-18666-5_5

available incidentally, but on a sound foundation that reasons the use of respective subindicators even though their influence in a cross-sectional context might be insignificant in select periods. Additionally, using established indicators in the construction of a composite indicator assures that the composite itself is well founded, theoretically or empirically speaking.

- Intertemporal and interregional comparability (consistency)

  Referring to the criticism voiced by Andersson and Heywood (2009) or Knack (2006) on the use and interpretability of the Corruption Perception Index (CPI) and other indices that are based on subjective evaluations, all subindicators need to be intertemporally and interregionally comparable. This quality assures statistical robustness. When, however, in the following chapters the dynamics (intertemporal comparability) of the composite indicator and its relation to other indicators are considered, intertemporal comparability becomes essential. The composite indicator needs to be interregionally comparable so that it can be suitably used as tool to compare countries' performance with respect to sustainable development and deduce respective policy recommendations.

- Statistical robustness

  The statistical robustness of the method used in constructing the composite indicator, as well as of the indicator itself, becomes particularly important if the indicator is to be included into an econometrical model, for example, when in a later chapter its correlation with third-party indicators of sustainability is considered.

  Statistical robustness can be assured by basing the construction of the indicator on empirical results and not on presuppositions about the methodology or weighting schemes. Additionally, the OECD guidebook on the construction of composite indicators (OECD 2008) has been considered in the construction of the indicator.

- Actionable

  All factors which are part of the composite indicator can be addressed directly and provide tangible for policy measures. Thus, it is possible to construct policy programs to counter problems in those areas where the composite indicator, or any of its subindicators, detects a shortcoming.

- Transferable and scalable

  Basically, the indicator does not discriminate according to the underlying geographical scope. It can, without loss of explanatory power, be applied on a national scale as well as on a regional or a global scale as long as the necessary data is available. Even a transfer to individual firms or financial products is possible.

- Durable

  The subindicators are relevant to the long-term development of a country as well as in the short term. In this context, an unchanged version of the indicator needs to remain relevant over the course of the next decades. In other words, the scope of the indicator is a sustainable long-term perspective instead of an environmentally advantageous positive short-term perspective.

- Relevant
  The indicator and all of its subindicators need to be useful to their target audience. This aspect necessitates a prior discussion of both the goal and target audience of the indicator as has been done in previous chapters. It needs to be made clear whether the indicator is confined to only one area or if it is a comprehensive indicator encompassing all important areas.
- Meaningful
  The indicator, and its constituent parts, must be understandable to the target audience. This includes a transparent methodology which implements indicators and concepts that can intuitively be grasped by the audience. Indirectly, this aspect necessitates that only a limited number of subindicators are implemented or else the indicator might become too confusing.
- Objective
  It needs to be possible for any third party to replicate parts of the indicator assuming they have access to the respective data sources. Additionally, the indicator needs to be based on facts and statistics, not on opinions or subjective evaluations. In this context, objectivity and intertemporal stability complement each other.
- Practical
  The composite indicator can be easily calculated and updated by implementing existing indicators and statistics. No additional costs are incurred for data generation or collection and both the subindicators and the indicator itself can be updated at regular intervals.

  Combining the aspects listed above leads to an indicator which is made up of only a select number of subindicators for which objective, quantitative data is easily and freely available but nonetheless which represents major dimensions of sustainability. Furthermore, each subindicator needs to be targetable by policymakers directly. Finally, the calculation scheme of the final indicator needs to be transparent, statistically stable, and founded by theoretical or empirical evidence.

  The presented EIIW-vita indicator fulfills all of these requirements—as will be shown on the next pages—and, while it is calculated solely on a national scale, it is scalable to other regional entities. Its basic concept can even be applied in the context of investment portfolios.

Selection of Subindicators

SEDAC (2007) already lists some 426 subindicators for measuring sustainability (including some duplicates as well) that are part of one of the six observed comprehensive indicators (Environmental Performance Index, Environmental Sustainability Index, Environmental Vulnerability Index, Rio to Johannesburg Dashboard of Sustainability, the Wellbeing of Nations, and the National Footprint Accounts). The UN (2007) and the European Commission (2009) list additional indicators and indicator groups. Consequently, evaluating all possible subindicators of sustainability is no simple task and neither is combining all of them into one composite indicator. Besides, using these indicators would not add any significantly new knowledge to the sustainability debate.

Even classifying the indicators into the three major groups of economic, environmental, and social subindicators, as used by the UN (2013), might not be suitable as, in the understanding of the authors of this study, the concept of sustainability is foremost an environmental and possibly economic concept and only secondly a social concept.

Furthermore, the use of too many subindicators would lead to a significant loss in practicality, meaningfulness and transferability of the indicator. Thus, combining all available subindicators into a single composite indicator does not provide a suitable solution.

A possible method of deducing meaningful subindicators from the pool of available subindicators would be to apply factor analysis to all of the available subindicators and, for each factor, to select one or two representatives which load highest on the respective factor.

However, the present study skips one step ahead and does not consider all subindicators, but only a preselection of four indicators (the selection of which can be justified):

- Genuine savings rate
- CO_2 emissions per unit of GDP
- Volume-weighted revealed comparative advantages (RCA) of the export of green products
- Comparative advantage of renewable energy

Firstly, the genuine savings rate, as published by the World Bank, accounts for savings (an economic concept of sustainability) but also includes natural resource exploitation and changes in the environment. Therefore, it is smaller than the base savings rate if natural resources are depleted faster than new resources are discovered or if the environment is harmed without providing compensation. However, it can also be larger than the original savings rate if new natural resources are discovered or fostered and the environment is strengthened, for example, via reforestation.

Secondly are the CO_2 emissions per unit of GDP. The amount of CO_2 emissions not only represents part of the conditions of the Kyoto protocol but also describes harm to the environment via a potential intensification of the greenhouse effect. Furthermore, this indicator allows for an assessment of the environment friendliness of the industry in an economy, as emissions are measured as a percentage of GDP and not in absolute terms. As the genuine savings rate already accounts for harm to the environment, the amount of CO_2 emissions might turn out to be superfluous as part of the final composite indicator.

Thirdly is a volume-weighted RCA (the traditional RCA is multiplied with the export volume to add an absolute sectoral dimension to the relative sectoral dimension of the RCA). The resulting indicator is then rescaled to the interval

[−1; 1], where the maximum value is assigned to 1 and the minimum value to −1). This not only allows accounting for the importance of green goods in the economy itself but also gives an idea about the standing of said economy in relation to all other economies in a reference market. Implementing the volume-weighted RCA therefore includes an economic dimension into the final indicator. However, it also has an environmental dimension, as it gives an insight into the economy's potential to be industrially environmentally friendly by producing environmentally friendly goods as well as by reporting on possible environmental spillovers into those countries which import the green goods.

Finally, the comparative advantage of renewable energy is more or less an addition to the previous subindicator. It shows to what extent, relative to other countries, green technologies—as in sources of renewable energy—are used in the economy.

A more in-depth analysis of the motivation behind the selection of these indicators in particular, out of the hundreds of possible indicators, can be found in the preceding chapters or indeed in Welfens et al. (2010).

Box 5.1 Idea of Revealed Comparative Advantages (RCA)

The idea of RCA follows the approach by Balassa (1965), while here a modified variant is used that relates to Borbely (2006) and is calculated as follows:

$$\mathrm{RCA}_{c,j} = \mathrm{tanhyp}\left(\ln\left(\frac{x_{c,j}}{\sum_{n=1}^{N} x_{c,n}} - \frac{\sum_{m=1}^{M} x_{m,j}}{\sum_{m=1}^{M}\sum_{n=1}^{N} x_{m,n}}\right)\right)$$

The indices c and j give the respective country c and the respective sector j, while M is the total number of countries and N is the total number of sectors. x represents the variable under consideration. Here it is either the export of green goods or the amount of renewable energy consumed. The tangens hyperbolicus and the natural logarithm are used to scale the indicator to the interval [−1; 1]. An RCA within the range of [−1; 0] describes a comparative disadvantage, while an RCA within the range of [0; 1] describes a comparative advantage. A larger absolute value signifies a larger disadvantage.

Calculation of the Composite Indicator

Referring to the OECD Handbook for Composite Indicators, it is noted that an indicator that is not evenly distributed—as are the three subindicators—can either be distributed across the whole interval, with the maximum values being set to unity and the minimum value being set to minus unity, or it can be evenly distributed with its mean fixed on zero.

In the construction of the EIIW-vita indicator, the second possibility is selected for two obvious reasons. Firstly, fixing the extremes of the indicator to the maximum and minimum of the series for each period would result in using a different distribution in each period. Secondly, using the second method, a clustering of value might occur which might lead—as is the case for some of the subindicators—to the result that only some individual countries report positive values, while the majority report negative values.

The most significant disadvantage of the first method lies in the fact that it is possible that the full range of [−1; 1] is not used. Basically, assuming an even distribution of values would lead to a distribution of values in the interval [−0.5; 0.5]. In all cases, even if assuming a very skewed distribution of values, results would appear in an interval with a length of one which might be situated more on the negative or the positive side of the interval [−1; 1]. In general, using the second method will result in smaller values which, however, will not bias the composite indicator as a whole.

Figures 5.1 and 5.2 illustrate the results of both methods when applied to the volume-weighted RCA of the export of green goods.

It can be seen from both figures that there are only a select few countries which specialize in the export of green exports and the trends are more or less the same. It becomes obvious that the second method discriminates better between the countries with an advantage and the countries without one. However, in general, it leads to much smaller indicator values than the first method which spans the whole range of indicator values.

In this study, the second method is implemented. It is realized by first subtracting the mean from each value of each period and then dividing the result by the range—the difference between the maximum and the minimum value of each period.

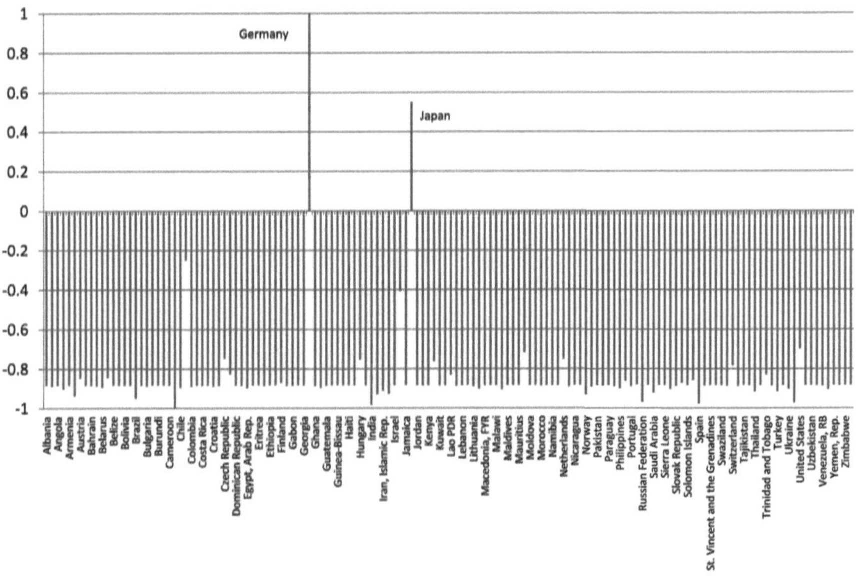

Fig. 5.1 Volume-weighted RCA for the export of green goods for selected countries in 2011—first method. *Source*: EIIW calculations based on data of World Bank, World Development Indicators and UN Comtrade, DESA/UNSD

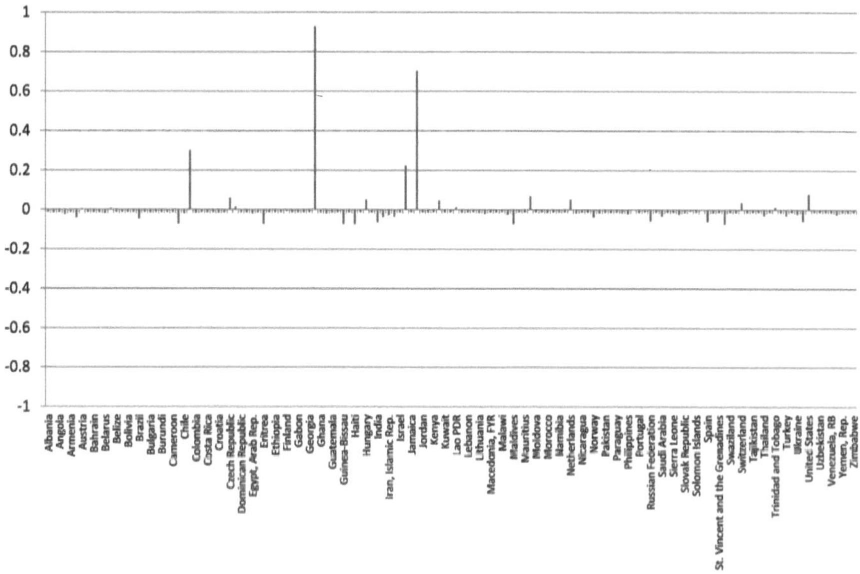

Fig. 5.2 Volume-weighted RCA for the export of green goods for selected countries in 2011—second method. *Source*: EIIW calculations based on data of World Bank, World Development Indicators and UN Comtrade, DESA/UNSD

In addition, Figs. 5.2, 5.3, and 5.4 give a rough preliminary impression about the distribution of the first and fourth selected partial indicators after centralization and standardization.

To check whether each of the four aforementioned indicators provides additional significance to a composite indicator, factor analysis has been used on all four indicators. As the goal has been to create only a single, one-dimensional indicator, the number of factors has been set to one.

In this context, the relevant result of the analysis is the eigenvalues. Traditional factor analysis would look for as few eigenvalues larger than unity as possible, as the number of these eigenvalues determines—if, for example, the Kaiser criterion is implemented—the number of factors. The ideal result would be for one eigenvalue to equal the number of subindicators—which gives both the maximum size of the eigenvalues and the sum of all eigenvalues—and the other eigenvalues to be zero or close to it. This situation would signify that all subindicators measure inherently the same concept, and therefore, only one significant factor exists.

In the context of this chapter, the question changes, however, to which indicators are dissimilar in the concepts they represent. Therefore, the ideal results would be for all indicators to be equal to unity, which would signify that each indicator describes an inherently different concept or in other words a different dimension of sustainability.

Applying factor analysis to the four partial indicators reveals that only three of the four indicators provide significantly different contexts. A correlation analysis then reveals that the CO_2 emissions in relation to GDP is superfluous in calculating

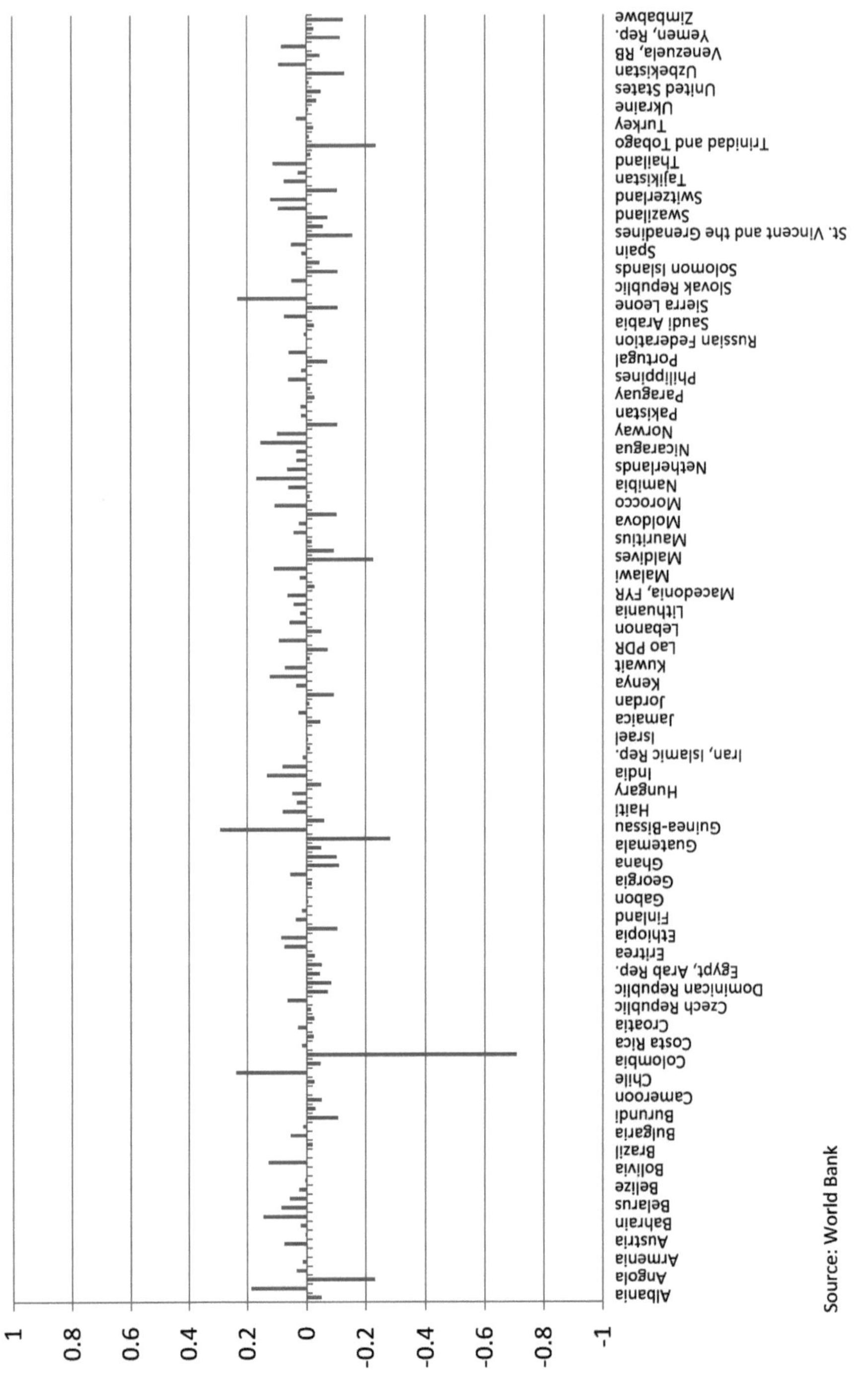

Source: World Bank

Fig. 5.3 Genuine savings rate for selected countries in 2011

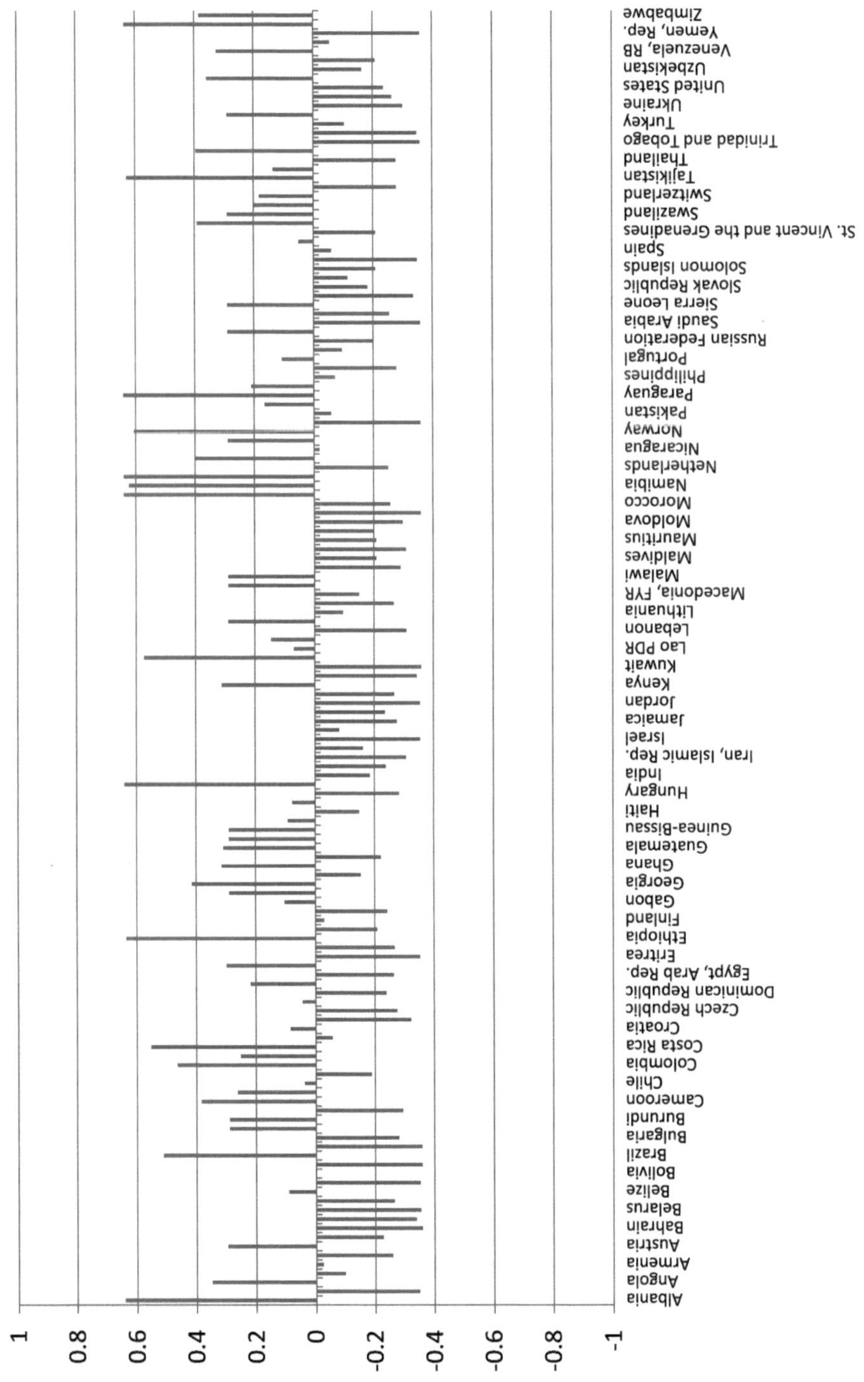

Fig. 5.4 RCA of the share of renewable energy for selected countries in 2011

Table 5.1 Weights determined via factor analysis for 3 selected years

	2000	2006	2007
Weighted RCAs of green goods	0.01	0.29	0.30
Genuine savings rate	0.50	0.39	0.38
Renewable RCA	0.50	0.32	0.31

a composite indicator, as its contents are already accounted for by the other three indicators, in particular the genuine savings rate.

Therefore, the set of subindicators is reduced to three and the emissions variable is excluded from further calculations.

Finally, the last step in calculating the EIIW-vita composite indicator would be to set the weights by which the three subindicators enter the composite indicator.

The first possibility lies in using exogenously set weights. In the simplest version, the composite indicator is the sum of each subindicator weighted by one third. Appendix J illustrates the calculation of the EIIW-vita indicator using a fixed weighting scheme.

Other externally motivated weights can prove helpful if specific aspects of the EIIW-vita indicator—specific dimensions of sustainability—should be emphasized.

Alternatively, the weights can be endogenously determined. In this case, weights would be deduced from the factor loadings yielded by the factor analysis.

Table 5.1 recaptures the results from Welfens et al. (2010) concerning weights determined via factor analysis for 3 selected years.

It can be seen, particularly in the later years, that the assumption of equal weights is no far-fetched assumption, whereas in 2000 the RCA of green goods is insignificant. This effect might be explained by a rise in environmental awareness during the course of the 2000s.

Estimation of Missing Values

For a large number of countries, data is only available on a sporadic basis. In these cases, missing data is approximated if possible by inter- or extrapolation of the existing data. In a few cases where this procedure is not possible, but where at least data for the genuine savings rate has been available, missing values were approximated. Country group-specific values have been used for the share of renewables. Table 5.2 gives an overview over the implemented estimates.

In the case of missing green exports, six strategies have been implemented. For Haiti, the amount of green exports has been set to zero. For both Tajikistan and Uzbekistan, the same values as those for Kazakhstan have been assumed. For Laos, an average of the other four Southeast Asian states has been calculated. For Angola and Sierra Leone, an average of all the other sub-Saharan countries has been calculated. For the Solomon Islands, an average of the other Melanesian states has been calculated. Finally, for all other cases of missing data, the growth rate that has been used for extrapolation has been estimated as the growth rate of the

Table 5.2 Replacement values for the share of renewables

Region	Countries	Value (%)	Reason
North African countries	Mali, Mauritania	5	Reference countries report values around 5 %
Sub-Saharan countries/South African countries	Burkina Faso	65	Reference countries report 50–80 %, while the group of sub-Saharan countries is set at 50 %. Energy use is mainly based on primary energy consumption, e.g., wood
	Burundi		
	Central African Republic		
	Chad		
	Equatorial Guinea		
	The Gambia		
	Guinea		
	Guinea-Bissau		
	Lesotho		
	Madagascar		
	Malawi		
	Niger		
	Rwanda		
	Sierra Leone		
	Suriname		
	Swaziland		
	Uganda		
Middle and South American countries	Belize	45	Reference countries report 40–50 %
	Guyana		
Islands	Comoros	15	Reference countries and group of Pacific Islands report 15 %
	Fiji		
	Maldives		
	Mauritius		
	Solomon Islands		
	St. Vincent and the G.		
Others	Djibouti	–	Similar to Yemen
	Bhutan	–	Similar to Nepal
	Lao PDR	–	Similar to Vietnam
	Papua New Guinea	–	Similar to Indonesia

previous period times the exports of the current period divided by the exports of the previous period.

If data for the genuine savings rate was missing, it has been inter- or extrapolated. In the case of Eritrea, the last reported value was from 2006 and it has been assumed to remain constant for the following years from 2007 to 2011. For Iran, the value for 2007 is set equal to the value for 2006. For Guinea-Bissau, Lesotho, and Uzbekistan, values for 2006, 2007, and 2008 are calculated as an average of the values for surrounding comparable countries, as was done in a previous paper by Welfens et al. (2010). These values are then used as a basis for additional intra- and extrapolation. For Zimbabwe, only a single savings rate has been reported; thus, it is assumed to remain constant for all periods.

Comparing the Findings of the Global Sustainability Indicator with Other Indicators

In previous chapters, it has been argued that a large number of indicators of sustainability exist and a discussion of this fact is no longer necessary at this point. It has been particularly stressed that the EIIW-vita indicator is only one among many composite indicators. Most of them have a specific area of interest and, in most cases, the indicators try to differentiate themselves from other indicators by targeting a new area of sustainability.

As ecology and the economy are intricate constructs of a myriad of entities interlinked with each other, it cannot be excluded that links between indicators of sustainability do exist. As all of them aim at the general topic of sustainability, it is even assumed a priori that links inherently must exist.

The goal of this chapter is to provide statistical results on the relationship between the EIIW-vita indicator and a number of other well-known indicators, for which in some cases a long-term comparison has even been possible.

The academic value of this analysis results from knowing which indicators are highly correlated to the EIIW-vita indicator and measure the same or related aspects of sustainability or the opposite, respectively.

If, however, no significant correlation with the EIIW-vita indicator exists, both indicators are independent of each other and the second indicator can be considered a suitable supplement for the EIIW-vita indicator.

The implemented indicators and available years were:

EIIW-vita indicator	2000–2011
Human Development Index	2000, 2005–2011
Corruption Perception Index	2000–2011
National ecological footprint	2006, 2007
Ecological footprint—biocapacity	2006, 2007
Environmental Performance Index	2000–2010
Moody's country rating	2007
Fitch country rating	2007
Standard and Poor's country rating	2007

While the first five indicators provide a numerical rating scale, a procedure has been used to code the three country ratings into numerical indices as well.

Although most indicators are based on different scales, a standardization is not necessary as the following analysis only relies on the ranking and not on the indicators' values.

Using this procedure is especially important because the construction of the indicator does not necessitate that the underlying scales are evenly distributed. It is also not possible to compare the scales from two different indicators to each other even if the indicators were to be standardized.

Preceding the analysis, it is important to get a deeper understanding of each of the indicators the EIIW-vita indicator is compared to.

Human Development Index (HDI)

The HDI is a composite indicator designed and published by the World Bank. While the basic structure of combining three partial indices—life expectancy, education, and income—remains the same across all years, the structure of the partial indicators as well as the method of their calculation changed in 2010, a fact which might lead to a break in an intertemporal correlation.

As the HDI is built on objective statistics, it can be considered objective. However, it centers on the living conditions in a country and excludes ecological aspects. Therefore, it actually measures the sustainability of the standard of living.

Corruption Perception Index (CPI)

The CPI is published by Transparency International. The index score is based on at least three out of 13 different surveys conducted by 12 research institutions. The indicator is built on surveys which only reflect the opinion of the questioned interviewees (who, however, are considered experts on specific countries).

The CPI is thus highly subjective in its nature and an intertemporal, as well as a cross-sectional, comparison of the CPI scores is rather problematic, as is argued by Andersson and Heywood (2009) and Knack (2006).

Aiming at corruption or rather the perception of corruption, the CPI focuses mostly on the sustainability of institutions.

National Ecological Footprint (NEF) and Biocapacity (BC)

The NEF and the BC are published by the Global Footprint Network. As only data for the years 2006 and 2007 has been freely available, only those 2 years were included in the analysis.

The NEF is built from six subindicators, all of them representing aspects of the ecological system. The NEF basically summarizes the harm done to different parts of the ecological system, while the BC—based on five subindicators—measures the opposite aspect. In contrast to the NEF, the BC gives the ecological potential a country has. Thus, both indicators are used to calculate an ecological surplus or deficit.

They are both considered separately, as the NEF can be seen as a sustainability indicator focusing on the harm done to the ecological system while the BC is an indicator focused on the present potential.

Environmental Performance Index (EPI)

The EPI, in some instances also referred to as the Yale indicator, is the successor of the environmental sustainability indicator developed by the Yale and Columbia Universities in cooperation with the World Economic Forum and the European Commission.

The indicator builds on two so-called objectives, the environmental health and the ecosystem vitality, which themselves are subdivided into eight policy categories and based on a total of 25 indicators. A number of these indicators are composite indicators.

Reducing the index to its most basic parts, it is based on a large number of objectively quantifiable indicators; however, the calculation of these in some parts is subjectively biased.

As its name already suggests, the EPI is mainly a sustainability indicator focused on environmental sustainability, which is underlined by the partial indicators considered.

Moody's, Fitch, and S&P Country Rating (MOODYS/FITCH/S&P)

All three of the aforementioned rating agencies publish country-at-risk ratings describing the credit worthiness of a country. While these ratings are based in many cases on hard objective data, in the end, it is a subjective act to assign a specific rating to a country. While certain data like inflation and interest rates surely enter the rating process, for outsiders, it is almost impossible to deduce any kind of coherent objective model underlying the rating process as a whole.

Assuming employees of rating agencies work according to a preset plan, it should be possible to compare ratings from the same rating agency both intertemporally and across countries even if a subjective aspect is present. However, comparing ratings from different agencies would not allow for this assumption, and thus, it is not possible to assume the objective comparability of the ratings.

All three ratings can be considered to be sustainability indicators as they give insights into the sustainability of a country's economic policy as well as the health of its financial markets.

These short summaries of the implemented indicators show that each of them represents different aspects of the rather broad concept of sustainability. Finding, therefore, a correlation between the EIIW-vita indicator and any of these indicator can show in which areas the EIIW-vita indicator is still limited and which areas of the concept of sustainability are already covered by it (as well as which established indicators might supplement it).

Implementing indicators that are uncorrelated to the EIIW-vita indicator would allow to enhance it by including new aspects. However, it should be considered—as stated earlier in this chapter—that each of the introduced indicators is itself a composite indicator; therefore, combining it with the EIIW-vita indicator will only lower the practicality as more subindicators will need to be included.

Concerning the correlation analysis itself, two approaches have been used. Firstly, where possible (HDI, CPI, and EPI), a basic system GMM panel estimator with fixed effects has been used. The respective results are presented in the second column of Table 5.3. Additionally, a Spearman's rank correlation coefficient has been calculated for the pooled data sets.

Secondly, Spearman's coefficients have been calculated for every year data has been available for each of the indicators. To get a better impression of whether the correlation effects remain stable over time, the results for the CPI, HDI, and EPI have been summarized in Table 5.3. Figure 5.5 plots the individual results along with the results for the pooled data set to account for breaks in the results.

Recapturing the results of Table 5.3, the HDI, CPI, and EPI in a panel estimation layout yield a significant negative relation between the EIIW-vita indicator and the HDI and CPI. The correlation with the EPI is positive, however, insignificant.

If, on the other hand, the data is pooled and a Spearman's rank correlation coefficient is calculated, the correlation between EIIW-vita and both the HDI and CPI remains significantly negative. The correlation with the EPI is again positive; however, it becomes significant.

Referring in a third step to correlations for individual years—again on the basis of a Spearman's rank correlation coefficient—the significant negative correlation between EIIW-vita, HDI, and CPI prevails; however, the CPI becomes insignificant in select periods, mostly the earlier ones. The EPI reports a negative correlation in the first 3 years and a positive correlation in the later years, all of which are insignificant. Figure 5.5 illustrates these developments.

Summarizing these results, the HDI and CPI are significantly negatively correlated to the EIIW-vita indicator independent of the method of measurement. Therefore, following the argumentation from above, both indicators would not offer any additional new insights were they to be added to the EIIW-vita indicator.

The EPI, on the other hand, reports only sporadic correlation in the case of pooled data; it can thus be assumed that no strong link between both indicators exists. The EPI can therefore offer a suitable addition to the EIIW-vita indicator. However, it needs to be considered that the EPI is built from a large number of statistics, so it is no surprise that it contains information that the EIIW-vita indicator lacks. Adding the EPI to the EIIW-vita indicator would actually lower the

Table 5.3 Correlation statistics for selected indicators—basic EIIW-vita indicator

	Total	Panel[a]	2000	2001	2002	2003	2004	2005	2006	2007	2008	2009	2010	2011
CPI	−0.2117	−0.0107	−0.1139	−0.2311	−0.1968	−0.1116	−0.1042	−0.1624	−0.1862	−0.1258	−0.1112	−0.1541	−0.1561	−0.1909
	(0.000)	(−9.16)	(0.318)	(0.033)	(0.056)	(0.244)	(0.266)	(0.084)	(0.048)	(0.185)	(0.239)	(0.097)	(0.090)	(0.023)
HDI	−0.4906	−0.3322	−0.4002					−0.3723	−0.395	−0.387	−0.3669	−0.4079	−0.39	−0.4756
	(0.000)	(−23.53)	(−0.000)					(−0.000)	(−0.000)	(−0.000)	(−0.000)	(−0.000)	(−0.000)	(−0.000)
EPI	0.0555	0.0002	−0.0684	−0.0725	0.1082	0.0873	0.0844	0.0478	0.0494	0.067	0.114	0.0637	0.0598	
	(0.045)	(0.52)	(0.549)	(0.510)	(0.296)	(0.363)	(0.368)	(0.613)	(0.603)	(0.481)	(0.227)	(0.495)	(0.518)	
NEF									−0.3507	−0.4281				
									(0.000)	(0.000)				
BC									−0.0371	−0.1200				
									(0.706)	(0.172)				
MOODYS										−0.0003				
										(0.980)				
FITCH										−0.0446				
										(0.706)				
SP										−0.0274				
										(0.817)				

[a]Value in brackets is the t-value of the coefficient; in the other case, the value in brackets is the possibility of the correlation being equal to zero
Source: Transparancy International, World Bank, IMF, Economic Footprint Network, Moody's Fitch, S&P, EIIW, own calculations

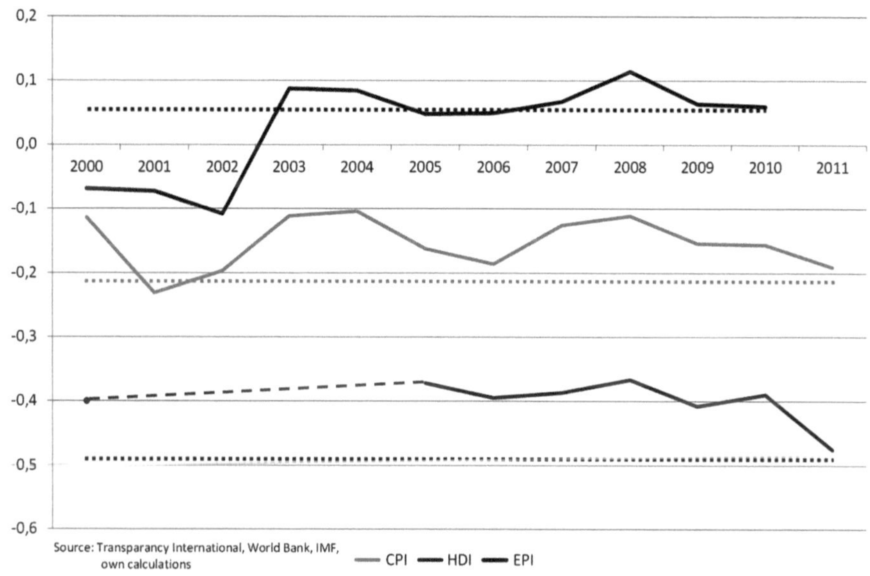

Fig. 5.5 Correlation dynamics for selected indicators

practicality of a resulting composite indicator. The EPI thus provides a possible however unsuitable addition to the EIIW-vita indicator.

Considering the NEF and the BC indicators, both report consistent negative correlations for each of the 2 available years. While the correlation with the NEF is highly significant, the correlation with BC is insignificant. The EIIW-vita indicator could therefore be enriched by adding a dimension of BC. As the NEF, similar to the genuine savings rate, focuses on the use of resources, its correlation to EIIW-vita is partially self-evident.

Finally, a financial market and governmental financial dimension is represented by the three country ratings. All three ratings report comparable negative correlations. In all three cases, the correlations were highly insignificant.

While the EIIW-vita indicator includes an economic aspect—via the green exports—it focuses only on parts of the economy. Country ratings can therefore offer a suitable addition to it. However, similar to the EPI, most country ratings are based on a large number of statistics and are also to be considered partially subjective. Therefore, an addition to the EIIW-vita indicator would limit both its practicality and objectivity at least.

If the EIIW-vita indicator is enriched by a dimension of water resource handling, it seems only prudent to recalculate the correlations and report on any significant changes.

While it has only been possible to calculate the extended EIIW-vita indicator for 3 years, it is an advantage that 2007 is one of those years, as the data for 2007 for all of the aforementioned and tested indicators has been available.

In contrast to the earlier analysis, only yearly correlations are calculated here.

Table 5.4 Correlation statistics for selected indicators—extended EIIW-vita indicators

	2002	2007	2011
CPI	0.0329	0.2836	−0.0796
	(0.752)	(0.017)	(0.348)
HDI		0.1476	−0.3722
		(0.223)	(0.000)
EPI	0.0329	0.3798	
	(0.504)	(0.001)	
NEF		0.0986	
		(0.147)	
BC		−0.0147	
		(0.904)	
MOODYS		−0.1916	
		(0.112)	
FITCH		−0.2376	
		(0.048)	
SP		−0.2081	
		(0.084)	

Value in brackets is the possibility of the correlation being equal to zero
Source: Transparancy International, World Bank, IMF, Economic Footprint Network, Moody's Fitch, S&P, EIIW, own calculations

Interpreting the results summarized in Table 5.4 shows that they are no longer as clear-cut as they have been before.

While the three country-at-risk ratings still remain negative, two of them become significant even though the EIIW-vita indicator has not been extended by any economic financial dimension.

It is furthermore important to note that the CPI and HDI report inconsistent results, while the EPI, however, still reports a positive correlation which becomes significant in 2007. The last result however is not surprising as the new dimension of the EIIW-vita indicator is also a significant aspect of the EPI.

While the BC remains insignificant, the NEF becomes insignificant as well which is counterintuitive, since previously it has been significantly correlated to the basic EIIW-vita indicator and the added dimension only accounts for one fifth of the extended indicator.

This part of the analysis is, however, flawed, as only 3 unconnected years have been available which means that for all indicators, with the exception of the CPI, only one or two observations were available, and thus, it is not possible to check whether the reported results are exceptions or if the reported results hold for all years and reflect on a general change in the correlations.

This problem already existed for the first part of the analysis; however, there it only involved the ratings and to a lesser degree the NEF and BC. The results there were, however, more coherent.

Nevertheless, in this chapter, it has been shown that the EIIW-vita indicator already accounts for different concepts of sustainability which were not initially part of its design, as it is highly correlated to a number of established sustainability indicators like the CPI, HDI, or NEF.

On the other hand, indicators have been identified which offer possible additions to the EIIW-vita indicator such as the EPI, BC, or country-at-risk ratings. However, the inclusion of any of these indicators is always accompanied by a loss in practicality and, in the case of the country-at-risk ratings, a loss of objectivity as well.

References

Andersson S, Heywood P (2009) The politics of perception: use and abuse of transparency international's approach to measuring corruption. Polit Stud 57:746–767

Balassa B (1965) Trade liberalization and revealed comparative advantage. Manch Sch 33:99–123

Borbely D (2006) Trade specialization in the enlarged European Union. Lucius & Lucius, Heidelberg

EPA (2012) A framework for sustainability indicators at EPA. EPA, Washington, DC

European Commission (2009) Sustainable development indicators. European Commission, Brussels

Knack S (2006) Measuring corruption in Eastern Europe and Central Asia. A critique of the crosscountry indicators. World Bank Policy Research Paper 3968

OECD (2008) Handbook on constructing composite indicators. OECD, Paris

SEDAC (2007) SEDAC compendium of environmental sustainability indicator collections version 1.1 data dictionary. SEDAC, New York

UN (2007) Indicators of sustainable development: guidelines and methodologies. UN, New York

UN (2013) Global sustainability development report. UN, New York

United Nations Commodity Trade Statistics Database, Department of Economic and Social Affairs/Statistics Division. http://comtrade.un.org/db/. Accessed 21 Jan 2015

Welfens PJJ, Perret JK, Erdem D (2010) Global economic sustainability indicator analysis and policy options for the Copenhagen process. IEEP 7:153–185

6 General Sustainability Trends for Key Countries in the World Economy

Covering global sustainability requires having a consistent composite indicator and focusing on the development of sustainability over time and across countries (as well as looking at the aggregate result for the world economy). The indicator considered is constructed, in the basic version, of three elements. An extended version, which is available for a shorter time period—there is a problem of data availability—additionally takes into account the role of water productivity. Subsequently, the dynamics of the EIIW-vita Global Sustainability Indicator is analyzed with a specific focus on leading OECD countries but also with a focus on future key economic players in the world economy, namely, Brazil, Russia, Indonesia, India, China, and South Africa (BRIICS). Taking a closer look at these countries (with a rather low income or medium per capita income), which are expected to show relatively high growth in the first half of the twenty-first century, is particularly interesting, not least because the expected economic catching-up process in itself raises some specific challenges with respect to sustainability.

In a separate chapter, we take a closer look at selected countries of the European Union, the USA, and Japan. The following countries have been chosen for the analyses: Germany, France, and the United Kingdom, as big European economies, along with Greece, Italy, and Spain as countries that were strongly affected by the economic crisis and face the challenge of adopting structural reforms as a means to improve the economic and social situation. Sustainability issues might not always be a top priority in countries that face serious macroeconomic adjustment challenges.

The Global Sustainability Indicator covers the period from 2000 to 2011. It allows us to see the dynamics in the development of the global sustainability in the beginning of the twenty-first century. The EIIW-vita Global Sustainability Indicator consists in the basic version of three sub-indicators:

- A savings sub-indicator measuring adjusted net savings, including particulate emission damage (% of GNI)

P.J.J. Welfens et al., *Towards Global Sustainability*,
DOI 10.1007/978-3-319-18666-5_6

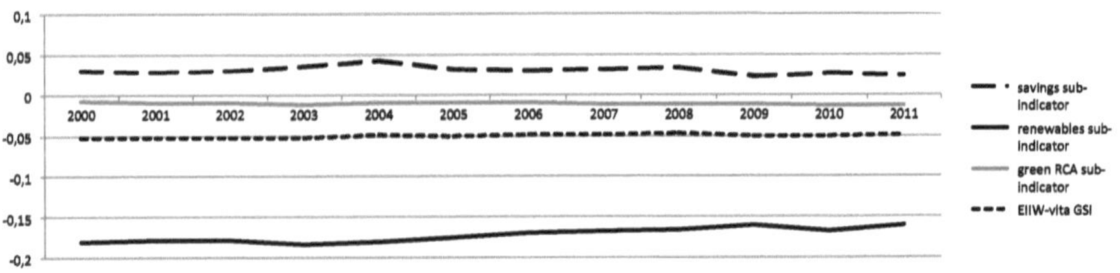

Fig. 6.1 Development of the World EIIW-vita Global Sustainability Indicator, 2000–2011. *Source*: EIIW calculations based on data of World Bank, World Development Indicators and UN Comtrade, DESA/UNSD

- A renewables sub-indicator measuring the percentage of the electricity produced from the renewable sources in total electricity production
- A Revealed Comparative Analysis (RCA) or "green competitiveness sub-indicator" that measures the specialization of the country on the exports of "environmentally friendly products"

The extended version of the indicator covers one additional dimension, namely, water productivity. The data for the calculation of the indicators are taken from the World Development Indicators database of the World Bank and from the UN Comtrade database.

In the subsequent graphical analysis for key countries of the world economy, the basic version of the indicator is presented; however, the recent extended version of the EIIW-vita Global Sustainability Indicator is also shown.

The development of the EIIW-vita Global Sustainability Indicator and its three inputs can be seen in Fig. 6.1. Let us recall that each sub-indicator is in the range between −1 and +1, so that for each country the composite indicator is also in the range of −1 to +1; an indicator value close to 0 can be considered as neutral.

The world indicator value has been negative during the whole period between 2000 and 2011, and one cannot see significant changes in the development of the indicator. In 2011, the EIIW-vita GSI takes the value of −0.0499. Interpreting the world indicator is not straightforward, as firstly, the indicator covers a short period of time and environmental economics normally needs long-term analyses, secondly, the world is heterogeneous and the dynamics in single countries should be taken into consideration, and thirdly, the world value has been used as a benchmark for the calculation of one sub-indicator, namely, RCAs. As will be presented later, some countries have improved their sustainability in the period analyzed whereas others have shown worsening performances.

Taking a closer look at the development of sub-indicators, one can see that the savings sub-indicator is positive but since 2004 shows a negative tendency. On the contrary, the renewables sub-indicator is negative but shows improvement. The "Green competitiveness" sub-indicator is negative but close to zero; it stays almost stable over the period analyzed.

A closer look at the EIIW-vita GSI and its sub-indicators can deliver a better understanding of the dynamics of the sustainability in the world. Firstly, the groupings representing the Top 20 and Bottom 20 countries are presented and discussed. Further, an extended version of the EIIW-vita GSI with the fourth sub-indicator, i.e., the water productivity sub-indicator, is shown and compared with the standard indicator. The values of each sub-indicator and the corresponding rankings are also presented. Finally, country profiles for the selected countries can also be found. For the study, the following countries are selected: fast-growing BRIICS countries (Brazil, the Russian Federation, India, Indonesia, China, and South Africa), selected European countries (France, Germany, the United Kingdom, Italy, Spain, and Greece), and the United States, as both one of the biggest economies in the world and one of the most important economic partners of the European Union.

Top and Bottom 20 Countries

Which countries have the best and the worst performance in sustainability according to the EIIW-vita GSI? Table 6.1 shows the results of the indicator with three inputs. The best performer in 2011 is Germany, followed by Nepal. Further European countries in the Top 20 group are: Norway in sixth place, Albania in tenth, and Iceland in eleventh. Six African, four Latin American countries, and three former Soviet Republics are also in the Top 20 list. Japan and New Zealand are also among the Top 20. As the Global Sustainability Indicator covers various dimensions of sustainability, the presence of such different countries among the top 20 is not surprising. Developed countries can naturally perform rather favorably in the export of green products, whereas developing countries could find it rather easy to be well positioned in the use of renewables. It is fair to say that the composite indicator has no bias for or against either developing countries or industrialized countries.

Among the Bottom 20 countries, there is one European country: Cyprus. Major exporters of natural resources, particularly of the oil and gas, are presented in the Bottom 20: Brunei, Syria, Yemen, Oman, and Saudi Arabia (see Table 6.1b). An important element for a rather unfavorable positioning lies in the poor genuine savings rate: those countries which strongly exploit their natural resources run the risk of a non-sustainable economic and ecological development, since no reinvestment in nonrenewable energy sources can be made. Non-sustainability as a serious problem might, however, only show up after several decades. Those countries in the Bottom 20 group should be encouraged to change their course of economic development and to adopt ecological and economic modernization on a broader scale. Here, the EIIW-vita Global Sustainability Indicator gives an important signal to policymakers. For investors, a poor position in the global ranking is a negative signal in the long run, and this should reinforce policymakers' willingness to switch to a more sustainable economic policy course in the future.

Comparing the Top 20 in 2000 and in 2011, we can see that some countries left the ranking while new leaders appeared (Table 6.1a, b). The countries that managed to improve their position in the EIIW-vita ranking are: Colombia, Georgia, Guinea-Bissau, New Zealand, Niger, and Germany, which moved from 24th place in 2000 to 1st in 2011.

Table 6.1 EIIW-vita GSI, Top 20 and Bottom 20

Top 20 countries	EIIW-vita GSI 2000	Ranking	Bottom 20 countries	EIIW-vita GSI 2000	Ranking
(a) EIIW-vita GSI in 2000 (three inputs)					
Namibia	0.2345	1	Yemen, Rep.	−0.1990	143
Japan	0.2262	2	Angola	−0.1846	142
Nepal	0.2224	3	Azerbaijan	−0.1672	141
Costa Rica	0.2165	4	Eritrea	−0.1602	140
Albania	0.2139	5	Brunei Darussalam	−0.1526	139
Norway	0.2136	6	Oman	−0.1503	138
Iceland	0.2134	7	Lebanon	−0.1487	137
Paraguay	0.2115	8	Bahrain	−0.1404	136
Mozambique	0.2056	9	Saudi Arabia	−0.1388	135
Ethiopia	0.1988	10	Trinidad and Tobago	−0.1311	134
Cameroon	0.1965	11	Syrian Arab Republic	−0.1304	133
Zambia	0.1843	12	Kazakhstan	−0.1281	132
Ghana	0.1823	13	Hungary	−0.1229	131
Uruguay	0.1796	14	Iran, Islamic Rep.	−0.1225	130
Tanzania	0.1766	15	Cambodia	−0.1216	129
Brazil	0.1728	16	Lithuania	−0.1209	128
Tajikistan	0.1696	17	Netherlands	−0.1199	127
Kyrgyz Republic	0.1563	18	Mongolia	−0.1191	126
Lesotho	0.1551	19	Algeria	−0.1188	125
Peru	0.1528	20	South Africa	−0.1180	124
(b) EIIW-vita GSI in 2011 (three inputs)					
Germany	0.2759	1	Trinidad and Tobago	−0.1942	143
Nepal	0.2653	2	Maldives	−0.1693	142
Ethiopia	0.2356	3	Yemen, Rep.	−0.1623	141
Tajikistan	0.2305	4	Oman	−0.1603	140
Namibia	0.2234	5	Mongolia	−0.1583	139
Norway	0.2226	6	Congo, Rep.	−0.1562	138
Mozambique	0.2050	7	Eritrea	−0.1514	137
Paraguay	0.1999	8	Saudi Arabia	−0.1393	136
Zambia	0.1995	9	Mali	−0.1384	135
Albania	0.1925	10	Syrian Arab Republic	−0.1329	134
Iceland	0.1924	11	South Africa	−0.1318	133
Costa Rica	0.1848	12	Brunei Darussalam	−0.1309	132
Kyrgyz Republic	0.1828	13	St. Vincent and the Grenadines	−0.1263	131
Guinea-Bissau	0.1702	14	Jordan	−0.1262	130

(continued)

Table 6.1 (continued)

Top 20 countries	EIIW-vita GSI 2000	Ranking	Bottom 20 countries	EIIW-vita GSI 2000	Ranking
Japan	0.1647	15	Kazakhstan	−0.1250	129
Brazil	0.1543	16	Lebanon	−0.1250	128
Niger	0.1442	17	Israel	−0.1247	127
New Zealand	0.1387	18	Tunisia	−0.1247	126
Colombia	0.1338	19	Cyprus	−0.1210	125
Georgia	0.1279	20	Benin	−0.1203	124
(c) Extended EIIW-vita GSI in 2011 (four inputs including water productivity sub-indicator)					
Germany	0.2197	1	Trinidad and Tobago	−0.1367	143
Singapore	0.2061	2	Yemen, Rep.	−0.1327	142
Nepal	0.1869	3	Mongolia	−0.1285	141
Norway	0.1836	4	Eritrea	−0.1254	140
Ethiopia	0.1654	5	Oman	−0.1238	139
Namibia	0.1634	6	Mali	−0.1158	138
Tajikistan	0.1606	7	Saudi Arabia	−0.1125	137
Iceland	0.1592	8	Syrian Arab Republic	−0.1115	136
Mozambique	0.1449	9	Kazakhstan	−0.1051	135
Paraguay	0.1439	10	South Africa	−0.1048	134
Zambia	0.1389	11	Tunisia	−0.1022	133
Albania	0.1336	12	Jordan	−0.1021	132
Costa Rica	0.1288	13	Lebanon	−0.0998	131
Japan	0.1248	14	Cambodia	−0.0956	130
Kyrgyz Republic	0.1248	15	Ukraine	−0.0946	129
Denmark	0.1233	16	Egypt, Arab Rep.	−0.0942	128
Switzerland	0.1169	17	Iran, Islamic Rep.	−0.0923	127
Guinea-Bissau	0.1164	18	Jamaica	−0.0922	126
Luxembourg	0.1118	19	Greece	−0.0919	125
Brazil	0.1085	20	St. Vincent and the Grenadines	−0.0915	124

Source: EIIW calculations based on data of World Bank, World Development Indicators and UN Comtrade, DESA/UNSD

The ranking for the extended version of the indicator, which additionally accounts for water productivity, is presented in Table 6.1c. The additional component slightly changes the position of countries in the ranking. Due to the high water productivity sub-indicator, the following European countries entered the Top 20 ranking: Denmark, Switzerland, and Luxembourg. The case of Singapore is notable: The country is the world leader as regards water productivity performance. That raises the country to second place in the extended EIIW-vita ranking.

New Dimension: Water Productivity

Water is an essential good for the survival of the mankind, it is extremely important for agriculture; it is used as an input in the production of other goods and as a mean of transportation. Hanemann (2006) underlined that water is a commodity with some special economic features. Firstly, water is both a private and public good. Secondly, water as a good is heterogeneous and has many dimensions such as quality, location, time, variability, and related uncertainty. The third feature is that the demand and supply structure of water is more complex compared to the other commodities, and the price building is different. Finally, water is an essential good for all life.

One can think that there is enough water on the planet for each of us. However, the problems of water availability and sustainable water use are becoming more and more important across the globe. Some countries have problems with access to water sources because of their geographical position and climate; other countries face the problem of water pollution because of industrial use or poor water management. It is clear that we cannot avoid the use of water by production and transportation, but it is in our hands to make it efficient and sustainable. The importance of the efficient use of water has been emphasized by the European Environment Agency (2012). One solution is the implementation of resource-efficient technologies which reduce the use of water and energy. Economic instruments like water pricing and market-based instruments can contribute greatly to the sustainable use of water resources.

The data on water use in Europe are presented in Fig. 6.2, where water productivity and water use per capita are presented, and Fig. 6.3 showing the data on water exploitation.

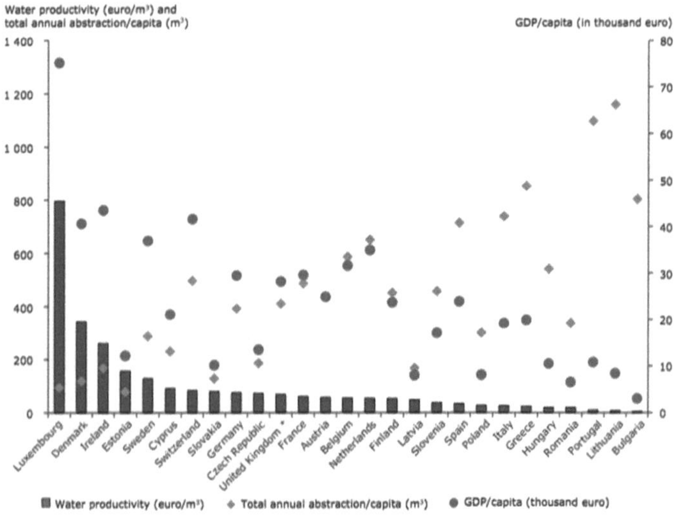

Fig. 6.2 Water productivity, economic output per capita, and water use per capita in EEA member countries. *Source*: European Environment Agency (2012)

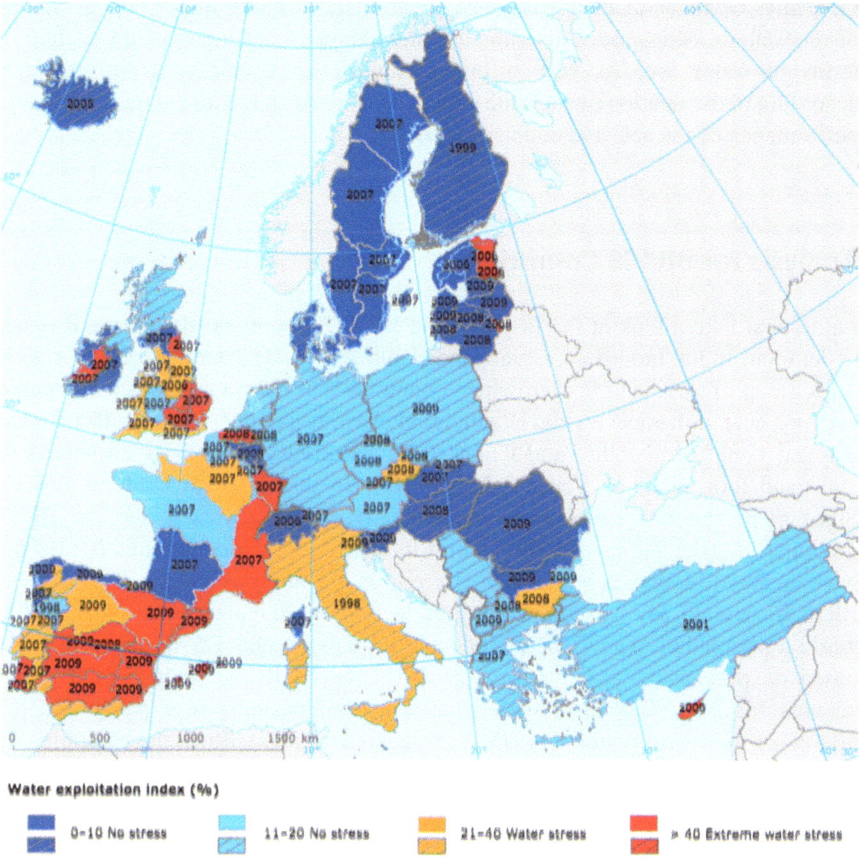

Fig. 6.3 Water exploitation index – towards a regionalized approach. *Source*: European Environment Agency (2012)

The basic EIIW-vita GSI was extended by one additional dimension to account for sustainable water use. A water productivity sub-indicator, calculated on the basis of the World Bank indicator which measures gross domestic product in US$ per cubic meter of total freshwater withdrawal, was used as an additional dimension.

The comparison between the basic and the extended version of the EIIW-vita GSI for the year 2011 is presented in Appendix G. The additional dimension has slightly changed the performance of the countries. The leader, according to the extended version, is Germany, followed by Singapore and Nepal in second and third place, respectively.

In Appendix H, one can see all four sub-indicators and the corresponding rankings. The best performers, according to the savings rate sub-indicator, are Guinea-Bissau, China, and Singapore. The best exporters of "green goods" are

Germany, Japan, and China. Paraguay, Iceland, and Albania are the best in the use of renewables. Singapore, Luxembourg, and Denmark are the most productive in terms of water use. As very different countries have the best performances according to the ranking of each sub-indicator, it is worth looking at the sustainable performance of the selected countries in detail.

Analysis for BRIICS Countries

Let us first take a look at the fast-growing BRIICS countries: Brazil, the Russian Federation, India, Indonesia, China, and South Africa. The acronym BRIC stands for a group of large and fast-growing economies that have a significant economic and political influence on development in their respective regions: Brazil, the Russian Federation, India, and China. The first BRIC forum was organized in 2009 and took place in Yekaterinburg, a city in the Southern Urals in Russia. Since then, summits have been held annually. South Africa joined the BRIC grouping in 2010 and hosted the summit in 2013 (Fifth Brics Summit 2014). We have included Indonesia in the BRICS group in our analysis and use the acronym BRIICS. Around 247 Million people live in Indonesia, representing 3.5 % of the world's population. The GDP of Indonesia equals around 1.5 % of world GDP based on purchasing power parity. In 2013, the BRIICS countries as a group account for 28.4 % of world GDP based on purchasing power parity and for 46.1 % of the world population (IMF Data and Statistics). Since the beginning of the twenty-first century, these countries have enjoyed high growth rates compared to the world average (see Fig. 6.4).

It is widely argued that fast-growing industrialized countries do not develop in a sustainable way, and that high growth rates cause pollution of the environment.

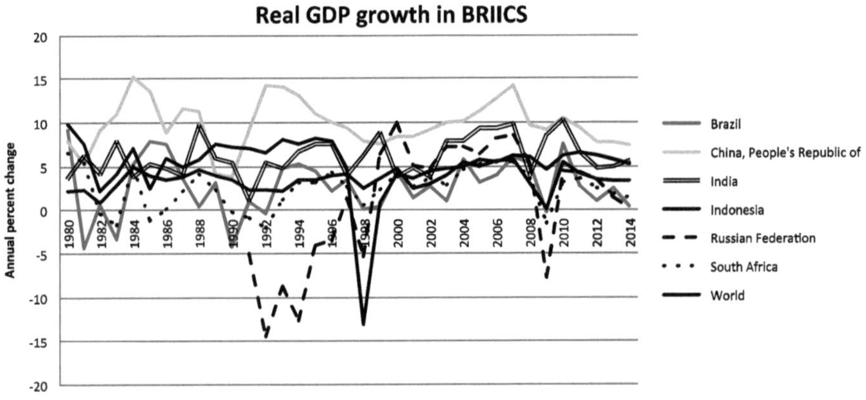

Fig. 6.4 Real GDP Growth in BRIICS, annual percent change. *Data Source*: IMF (2014) World Economic Outlook

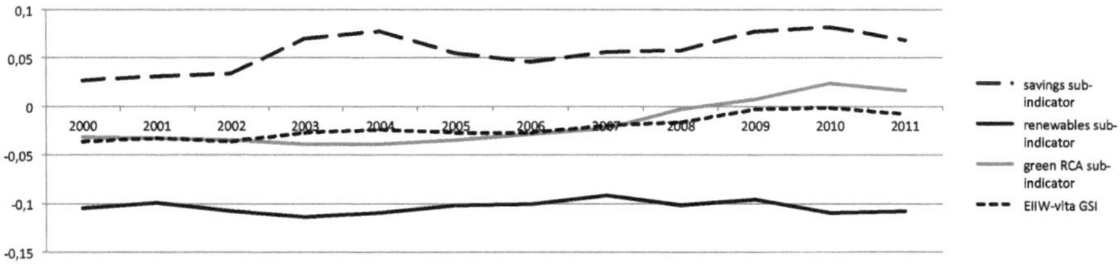

Fig. 6.5 Development of the EIIW-vita Global Sustainability Indicator and four Sub-indicators for the Group of BRIICS countries, 2000–2011. *Source*: EIIW calculations based on data of World Bank, World Development Indicators and UN Comtrade, DESA/UNSD

On the contrary, one can see that the EIIW-vita GSI for BRIICS has actually improved over the period between 2000 and 2011.

The savings sub-indicator shows positive dynamics from 2000 until 2004 and from 2006 until 2010. It is also worth mentioning that the sub-indicators are relative and the values change with the changes of the world performance. One can clearly see the improvement of the "green competitiveness" of the BRIICS countries, from being negative until 2007 it switched to a positive value from 2008 onwards. This can be primarily explained by the very good performance of China. The renewables sub-indicator shows negative dynamics (see Fig. 6.5).

Further, the country profiles are presented; countries are compared with the middle value of the BRIICS countries as well as with the world performance. The reader should pay attention to the scale of the diagrams shown below, as it differs for different countries so that the reader can clearly see the development of each sub-indicator and the composite EIIW-vita indicator. It is not straightforward to interpret the development of each indicator across time as negative dynamics in one country can be explained by not only the worse performance of the country itself but also by an improvement in other countries. The composite indicator and its sub-indicators are relative. For example, a significant improvement in terms of exports of "green products" in China has worsened the performance of other countries in the EIIW-vita ranking even though the exports of these countries continued to grow. The growth of "green exports" in these countries has just not been that big as in China. All the statistics are presented for the year 2011.

Country Profile Brazil 2011

GDP (constant 2005 US$)**[a] 1,126,722,915,143.3 **GDP per capita (constant 2005 US$)[a] 5,721.3	**Adjusted net savings, incl. particulate emission damage (% of GNI)**[b] 6.79 **Savings sub-indicator**[b] −0.0018 (75)
EIIW-vita GSI[b] 0.1543 (16 out of 143) **Extended EIIW-vita GSI**[b] **incl. water productivity** 0.1085 (20 out of 143)	**Green exports (current US$)**[b] 5,456,570,591 **RCA sub-indicator**[b] −0.0474 (133)
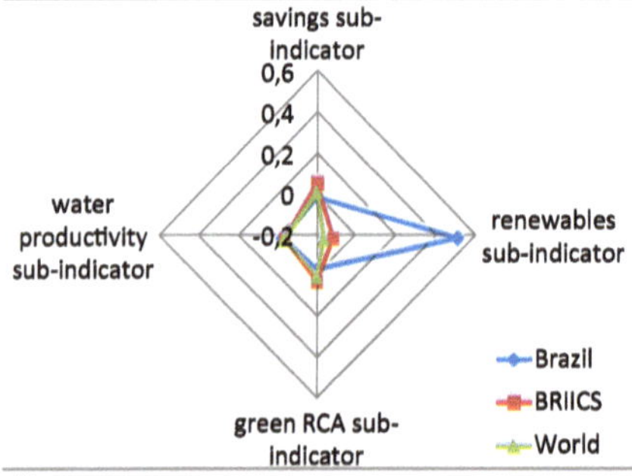	**Share of renewable energy sources in electricity production (% of total)**[b] 87.12 **Renewables sub-indicator**[b] 0.5121 (13) **Water productivity (constant 2005 US$ GDP per m³ of total freshwater withdrawal)**[b] 19.40 **Water productivity sub-indicator**[b] −0.0287 (61)

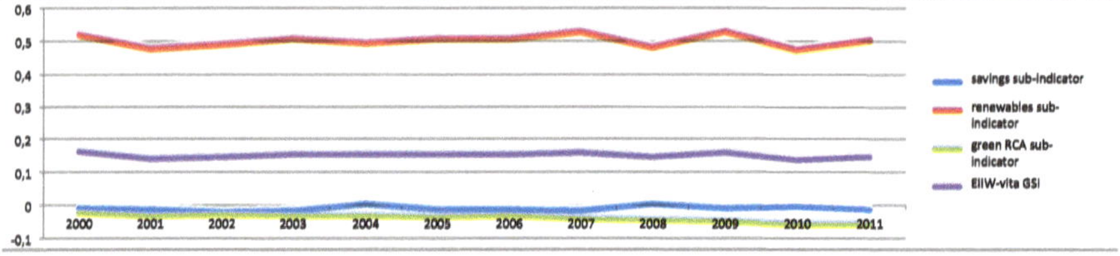

Note: Ranking position is indicated in brackets. The results are presented for the year 2011.
[a]World Bank, World Development Indicators.
[b]EIIW calculations based on data of World Bank, World Development Indicators and UN Comtrade, DESA/UNSD

According to the EIIW-vita GSI, Brazil is in 16th place out of 143 in 2011. As regards the savings and "green competitiveness" sub-indicators, Brazil is slightly below the world average and the average of BRIICS countries. Brazil has a very good performance in the use of the renewables. In 2011, 87 % of electricity in the country was produced from renewable sources. The main renewable source of energy is hydropower, followed by thermal power. The share of wind energy is still relatively small, but it is increasing over time (Brazilian Electricity Regulatory Agency 2013).

Country Profile Russian Federation 2011

GDP (constant 2005 US$) 948,013,362,239.7 **GDP per capita (constant 2005 US$)** 6,631.5	**Adjusted net savings, incl. particulate emission damage (% of GNI)** 7.16 **Savings sub-indicator** 0.0012 (72)
EIIW-vita GSI −0.0858 (106 out of 143) **Extended EIIW-vita GSI incl. water productivity** −0.0729 (110 out of 143)	**Green exports (current US$)** 3,665,973,663 **RCA sub-indicator** −0.0574 (134)

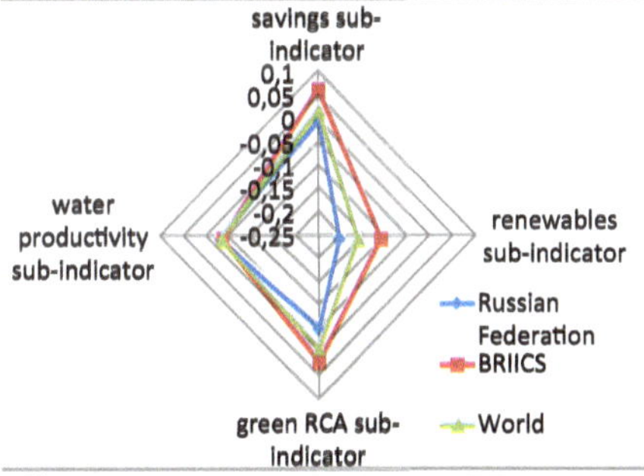

Share of renewable energy sources in electricity production (% of total)
15.80
Renewables sub-indicator
−0.2011 (85)
Water productivity (constant 2005 US$ GDP per m^3 of total freshwater withdrawal)
14.32
Water productivity sub-indicator
−0.0342 (71)

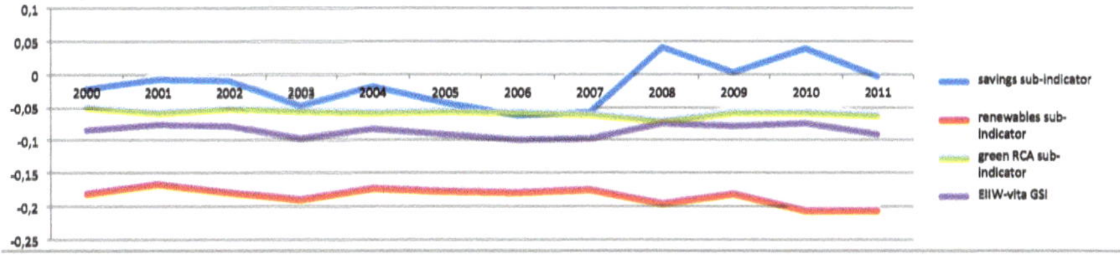

The Russian Federation emerged in 106th place according to the EIIW-vita GSI in 2011. The performance of the country in each single sub-indicator, except for water productivity, is lower than both the world and BRIICS averages. One can see a positive development as regards the savings sub-indicator and a negative development of the renewables sub-indicator after 2007. According to the World Development Indicators Database of the World Bank, 49.3 % of electricity in 2011 is produced from gas. The share of coal is around 15.5 %. The share of electricity produced from nuclear sources in the total production is around 16.4 %. The country produces 16 % of electricity from renewable sources, mostly from hydropower. The share of renewables in electricity production excluding hydropower is only 0.1 %, although the potential of using other renewable sources such as wind or solar power is rather high. The topic is gaining in importance, and there are several initiatives in this field.

Country Profile India 2011

GDP (constant 2005 US$) 1,326,235,203,611.8 **GDP per capita (constant 2005 US$)** 1,086.0	**Adjusted net savings, incl. particulate emission damage (% of GNI)** 23.49 **Savings sub-indicator** 0.1339 (8)
EIIW-vita GSI −0.0383 (79 out of 143) **Extended EIIW-vita GSI incl. water productivity** −0.0406 (85 out of 143)	**Green exports (current US$)** 5,029,999,580 **RCA sub-indicator** −0.0638 (137)
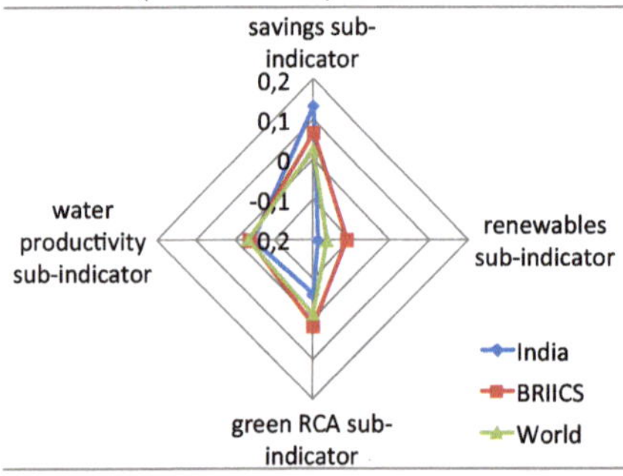	**Share of renewable energy sources in electricity production (% of total)** 17.42 **Renewables sub-indicator** −0.1849 (82) **Water productivity (constant 2005 US$ GDP per m^3 of total freshwater withdrawal)** 1.74 **Water productivity sub-indicator** −0.0476 (129)

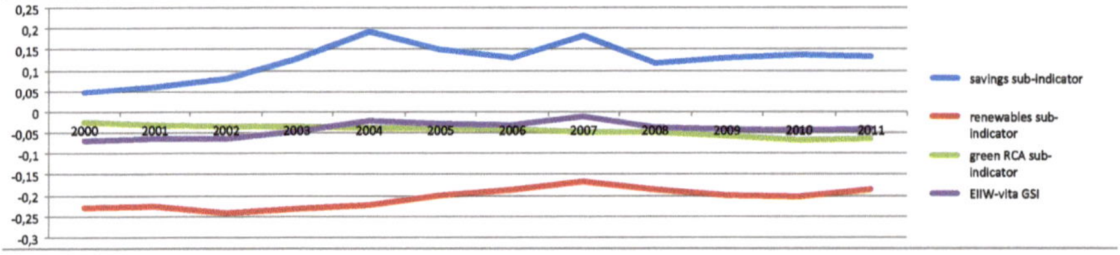

India demonstrates good performance in savings: India is in eighth place, with an adjusted net savings rate of 23.5 % of GNI in 2011. The green RCA and the renewables sub-indicators are negative. In India, around 67.9 % of electricity is produced from coal. The second important electricity source is hydropower with the share of 12.4 %. Electricity production from renewable sources other than hydropower accounts for 5 %. EY (2013) names six main reasons for the promotion of renewables in the country: energy security concerns, climate change, the increasing cost competitiveness of renewable energy technology, distributed electricity demand, government support, favorable foreign investment policy, and a vast untapped potential. The estimated potential of wind energy is 102.8 GW, of bio-power 22.5 GW and of solar power 6 GW. The composite EIIW-vita indicator for India is −0.0383, which puts India in 79th place.

Country Profile Indonesia 2011

GDP (constant 2005 US$) 402,408,024,159.8 **GDP per capita (constant 2005 US$)** 1,650.6	**Adjusted net savings, incl. particulate emission damage (% of GNI)** 17.05 **Savings sub-indicator** 0.0816 (22)
EIIW-vita GSI −0.0649 (90 out of 143) **Extended EIIW-vita GSI incl. water productivity** −0.0601 (101 out of 143)	**Green exports (current US$)** 2,329,712,483 **RCA sub-indicator** −0.0377 (131)

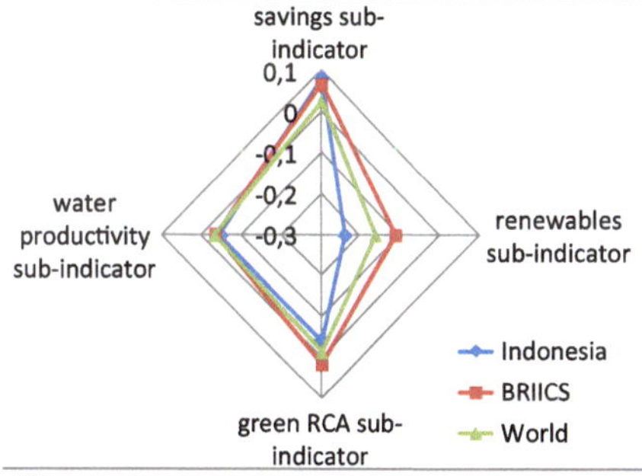

Share of renewable energy sources in electricity production (% of total)
12.06
Renewables sub-indicator
−0.2385 (96)
Water productivity (constant 2005 US$ GDP per m^3 of total freshwater withdrawal)
3.55
Water productivity sub-indicator
−0.0456 (115)

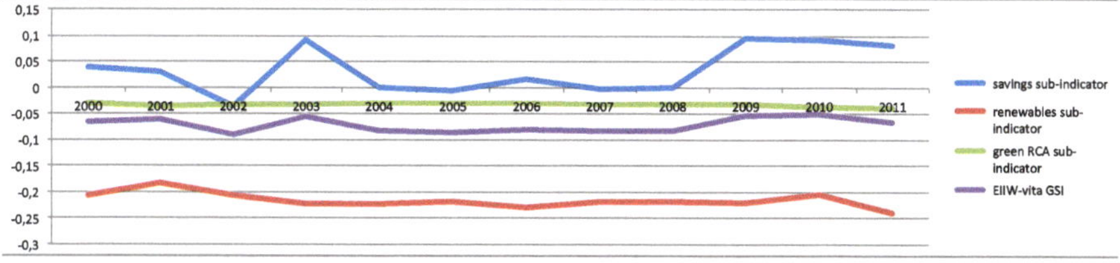

Indonesia occupies 90th place in the EIIW-vita ranking. The country performs well in savings but weakly in the renewables sub-indicator. In 2011, adjusted net savings was 17.05 % of the GNI. The share of renewables in electricity production in 2011 is 12.06 %. 44.4 % of the electricity is produced from coal, 23.2 % from oil, and 20.3 % from gas. Renewable energy is one of the main focuses of the Indonesian government. This focus is reflected in the Indonesian Energy Blueprint 2006–2025 (Ministry of Energy and Natural Resources 2006), which sets milestones for the development of renewable energy. The government has also issued Indonesian Presidential Regulation No. 5/2006, which stated that the share of renewable energy in the Indonesian energy mix should reach 17 % by 2025 (Government of Indonesia 2006). Issues which hinder the development of renewable energy are a lack of regulation, technology, and infrastructure.

Country Profile China 2011

GDP (constant 2005 US$) 4,196,333,171,021.3 **GDP per capita (constant 2005 US$)** 3,122.0	**Adjusted net savings, incl. particulate emission damage (% of GNI)** 36.44 **Savings sub-indicator** 0.2392 (2)
EIIW-vita GSI 0.1174 (23 out of 143) **Extended EIIW-vita GSI incl. water productivity** 0.0777 (29 out of 143)	**Green exports (current US$)** 93,318,621,342 **RCA sub-indicator** 0.3016 (3)

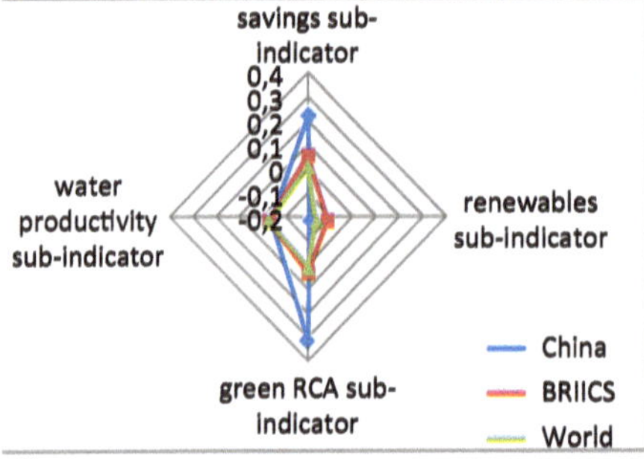

Share of renewable energy sources in electricity production (% of total)
17.04
Renewables sub-indicator
−0.1887 (83)
Water productivity (constant 2005 US$ GDP per m^3 of total freshwater withdrawal)
7.57
Water productivity sub-indicator
−0.0414 (90)

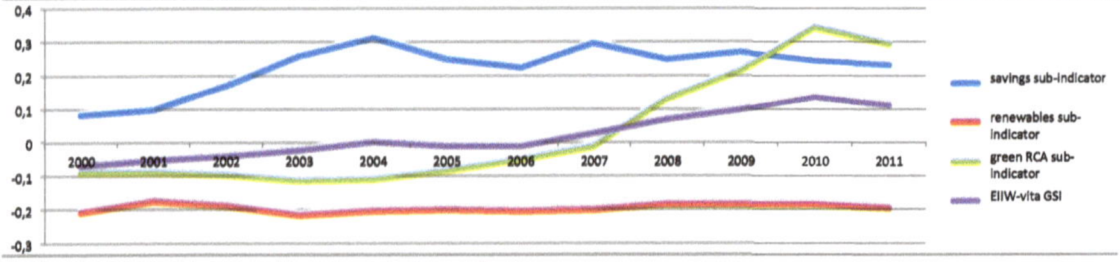

China is in second place worldwide in terms of savings and in third in terms of the export of "green products." Until 2007, the value of the "green exports" sub-indicator was negative; from 2007 one can see a significant rise in Chinese "green exports." In the period between 2000 and 2011, the "green exports" of China grew by a factor of more than 16. In 2011, China is the second biggest exporter of "green products" in absolute terms worldwide, after Germany. As regards the use of renewables in electricity production, China is below both the world and BRICS averages. There is a lot of potential for improvement in this field. In China, around 79 % of electricity is produced from coal, renewables are the second most important source with the share of 17.04, 14.82 % is produced with hydroelectric sources, and 2.22 % from the other renewable sources. The initiatives of the Chinese government in the field of environmental policy are presented in the White Paper on Energy Policy (State Council Information Office of the People's Republic of China 2012).

Country Profile South Africa 2011

GDP (constant 2005 US$) 300,242,871,561.1 **GDP per capita (constant 2005 US$)** 5,821.0	**Adjusted net savings, incl. particulate emission damage (% of GNI)** 1.55 **Savings sub-indicator** −0.0444 (105)
EIIW-vita GSI −0.1318 (133 out of 143) **Extended EIIW-vita GSI incl. water productivity** −0.1048 (134 out of 143)	**Green exports (current US$)** 4,535,316,821 **RCA sub-indicator** −0.0012 (17)
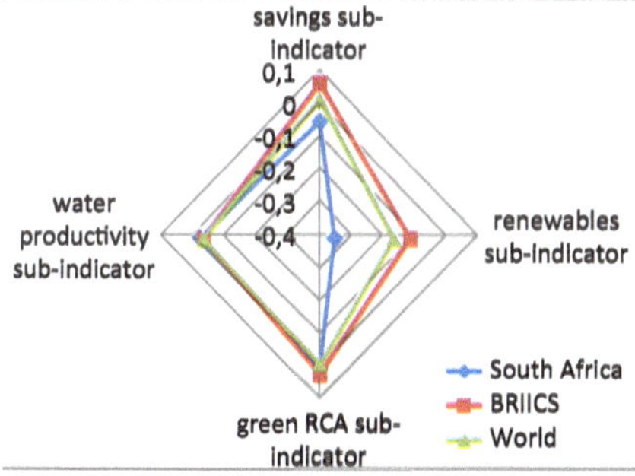 	**Share of renewable energy sources in electricity production (% of total)** 0.95 **Renewables sub-indicator** −0.3496 (129) **Water productivity (constant 2005 US$ GDP per m^3 of total freshwater withdrawal)** 23.97 **Water productivity sub-indicator** −0.0239 (52)

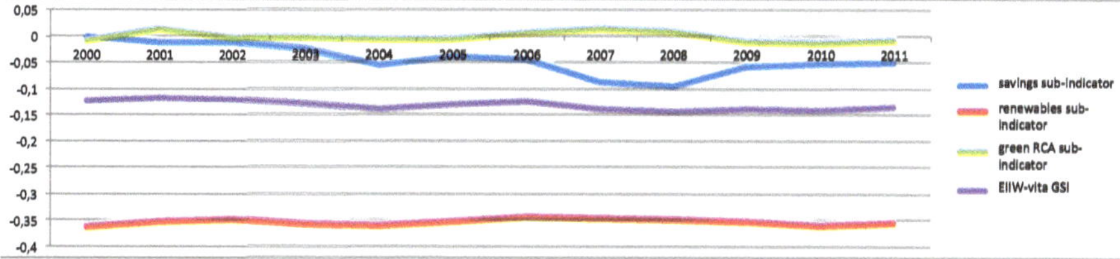

South Africa performs pretty poorly compared to the other BRIICS countries. The country lies in 133rd place as regards total performance, which puts the country in the Bottom 20 group. South Africa performs well in the exports of "green products," but as regards savings rate and renewables, the country has a low position in the ranking. Adjusted net savings equals only 1.55 % of the GNI in 2011. The share of renewable sources in electricity production is very low: only 0.95 %. Almost all electricity in the country (93.87 %) is produced from coal and 5.2 % from nuclear sources although wind and solar power have great potential in the country. The importance of the development of renewable energy has been underlined in the Integrated Resource Electricity Plan 2010–2030 (Government of South Africa 2011). The Department of Energy (2013) aims to improve the energy mix by having 30 % of clean energy by 2025, achieved by developing effective legislation and policies, encouraging investment, and diversifying in terms of the energy mix.

Analysis for the Selected European Countries

Now let us take the European perspective. Detailed analysis is presented for the following countries: Germany, France, the United Kingdom, Greece, Italy, and Spain. The case of Greece, as a Euro Crisis country, is of special interest for the analysis. The countries are compared with the middle value of the European Union, exclusive of Malta, as well as with the world performance. Malta is not covered by the study because of the unavailability of data which is why an EU 26 is used as a benchmark.

One can see from Fig. 6.6 an improvement in the use of renewables in electricity production since 2003. One can recall the Lisbon Strategy of the European Union from the year 2000, which aimed to provide sustainable development in the European Union until 2010. The share of renewables was one of the main indicators used to measure the success of the implementation of the Strategy. The savings sub-indicator is positive. However, the savings sub-indicator falls in 2009 because of the economic crisis but recovers in 2010 and 2011. The RCAs sub-indicator is positive over the whole period observed. The main drivers of European "green competitiveness" are Germany and Italy.

Two further countries are also analyzed: the United States and Japan. The United States is the largest economy in the world, an important trade partner for European countries and a big polluter. Japan is an important global player and a leading producer of high-tech innovative products, including "green products."

Please note that the scale of the diagrams shown below differs for different countries, to make it easier to recognize the development of each sub-indicator and the composite EIIW-vita indicator. All statistics presented are for the year 2011.

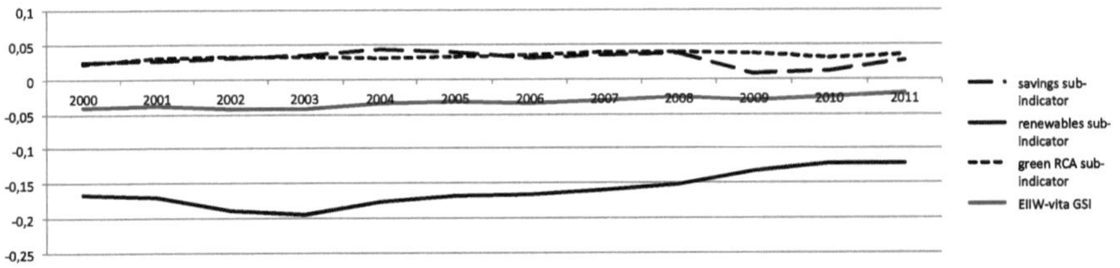

Fig. 6.6 Development of the EIIW-vita Global Sustainability Indicator and four Sub-indicators for EU 28 exclusive Malta, 2000–2011. *Source*: EIIW calculations based on data of World Bank, World Development Indicators and UN Comtrade, DESA/UNSD

Country Profile Germany 2011

<table>
<tr><td>GDP (constant 2005 US$)
3,052,837,685,917.5
GDP per capita (constant 2005 US$)
37,321.8</td><td>Adjusted net savings, incl. particulate emission damage (% of GNI)
13.79
Savings sub-indicator
0.0550 (37)</td></tr>
<tr><td>EIIW-vita GSI
0.2759 (1 out of 143)
Extended EIIW-vita GSI incl. water productivity
0.2197 (1 out of 143)</td><td>Green exports (current US$)
116,544,962,290
RCA sub-indicator
0.9273 (1)</td></tr>
<tr><td>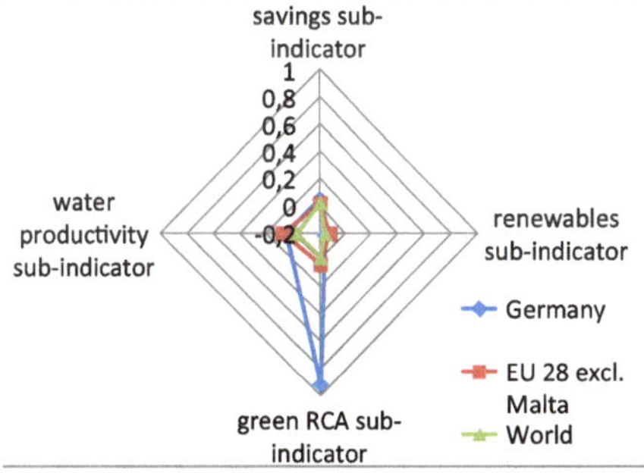
</td><td>Share of renewable energy sources in electricity production (% of total)
20.44
Renewables sub-indicator
−0.1547 (78)
Water productivity (constant 2005 US$ GDP per m^3 of total freshwater withdrawal)
94.39
Water productivity sub-indicator
0.0512 (17)</td></tr>
<tr><td colspan="2">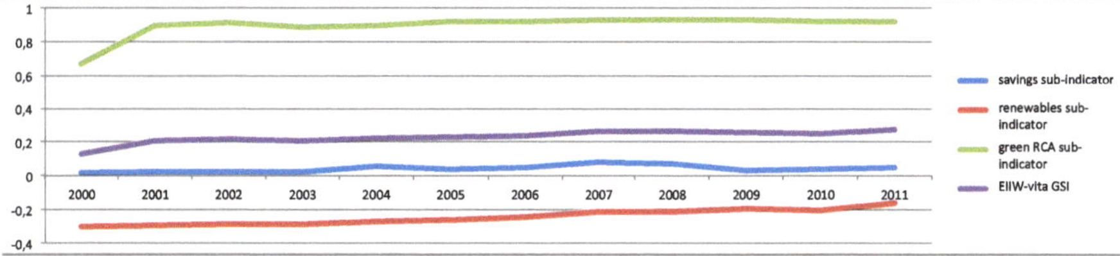
</td></tr>
</table>

Germany is the world leader in terms of "green exports"; this allows the country to achieve first place in the overall ranking. In 2011, the country exports "green products" to the value of US$116.5 billion. The share of "green exports" in the total exports in 2011 is 6.5 %. Compared to 2000 the "green exports" in 2011 have grown by a factor of circa 3.6. The renewables sub-indicator has a positive tendency but is still below 0. In 2011, Germany produced 20 % of electricity from renewable sources. 45.13 % of electricity is produced from coal, 17.92 % from nuclear sources, 13.88 % from natural gas, and 1.10 % from oil. The German Renewable Energy Act plays an important role in the promotion of the use of renewables (Ministry of Economy and Energy of Germany 2014). As regards the savings sub-indicator, Germany is ranked in 37th place, with a share of adjusted net savings in GNI to value of 13.79 %.

Country Profile France 2011

<table>
<tr>
<td>GDP (constant 2005 US$)
2,249,127,707,682.9
GDP per capita (constant 2005 US$)
34,420.0</td>
<td>Adjusted net savings, incl. particulate emission damage (% of GNI)
8.98
Savings sub-indicator
0.0160 (64)</td>
</tr>
<tr>
<td>EIIW-vita GSI
−0.0805 (102 out of 143)
Extended EIIW-vita GSI incl. water productivity
−0.0538 (95 out of 143)</td>
<td>Green exports (current US$)
23,180,732,785
RCA sub-indicator
−0.0148 (90)</td>
</tr>
<tr>
<td>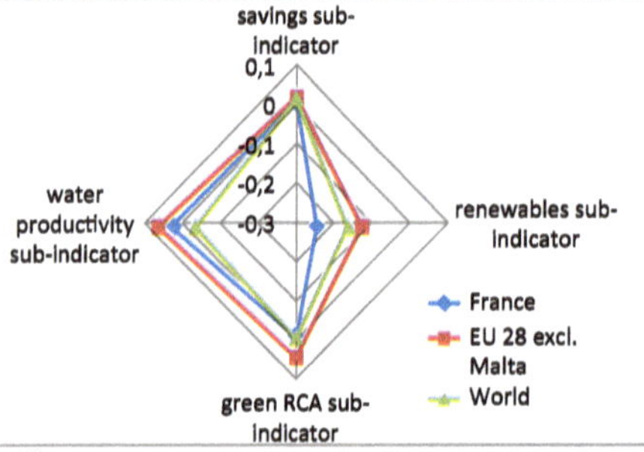
</td>
<td>Share of renewable energy sources in electricity production (% of total)
11.63
Renewables sub-indicator
−0.2428 (98)
Water productivity (constant 2005 US$ GDP per m³ of total freshwater withdrawal)
71.13
Water productivity sub-indicator
0.0264 (25)</td>
</tr>
</table>

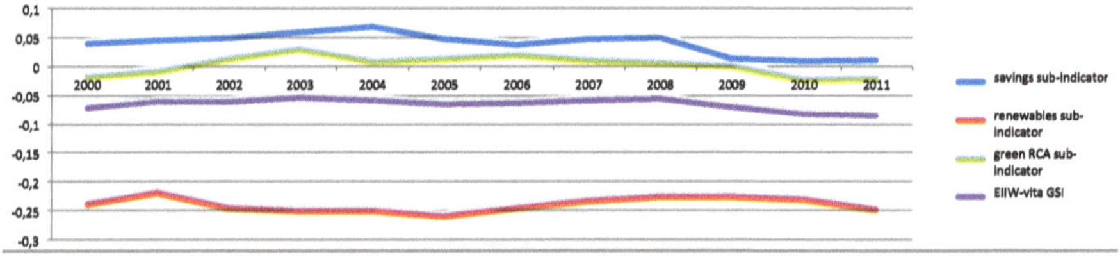

France is in 102nd place in the EIIW-vita GSI ranking. The main reason being the relatively small share of renewables in electricity production: around 11.6 %. In 2011, most of the energy in France is produced from nuclear sources (around 79.44 %). Among the most recent policy initiatives as regards the use of renewables are the National Renewable Energy Action Plan (NREAP), the Green innovation funding program from 2010, and the Renewable Energy Feed-In Tariff and offshore wind tendering mechanism (IEA 2014). Tax reduction is an additional instrument. Investors in plants that produce electricity from renewable sources can get an income tax credit. Those who install photovoltaic installations on buildings can apply for a reduction in the VAT rate (Res Legal 2014). The "green competitiveness" indicator of France is below the EU average.

Country Profile United Kingdom 2011

<table>
<tr>
<td>GDP (constant 2005 US$)
2,386,404,322,586.7
GDP per capita (constant 2005 US$)
37,724.4</td>
<td>Adjusted net savings, incl. particulate emission damage (% of GNI)
2.71
Savings sub-indicator
−0.0350 (104)</td>
</tr>
<tr>
<td>EIIW-vita GSI
−0.1196 (123 out of 143)
Extended EIIW-vita GSI incl. water productivity
−0.0531 (93 out of 143)</td>
<td>Green exports (current US$)
20,896,837,166
RCA sub-indicator
−0.0590 (135)</td>
</tr>
<tr>
<td>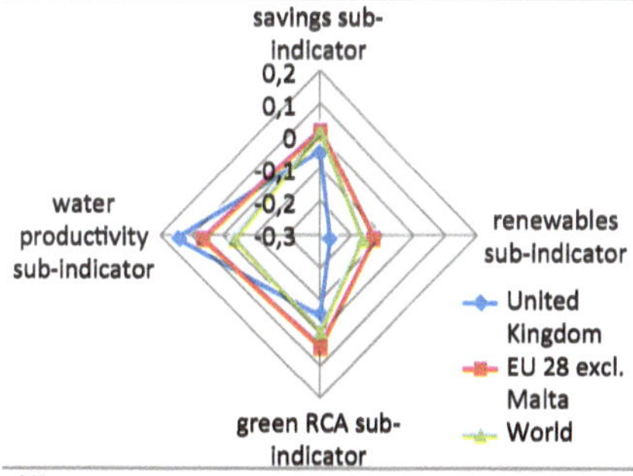
</td>
<td>Share of renewable energy sources in electricity production (% of total)
9.43
Renewables sub-indicator
−0.2648 (104)
Water productivity (constant 2005 US$ GDP per m^3 of total freshwater withdrawal)
183.73
Water productivity sub-indicator
0.1465 (6)</td>
</tr>
<tr>
<td colspan="2">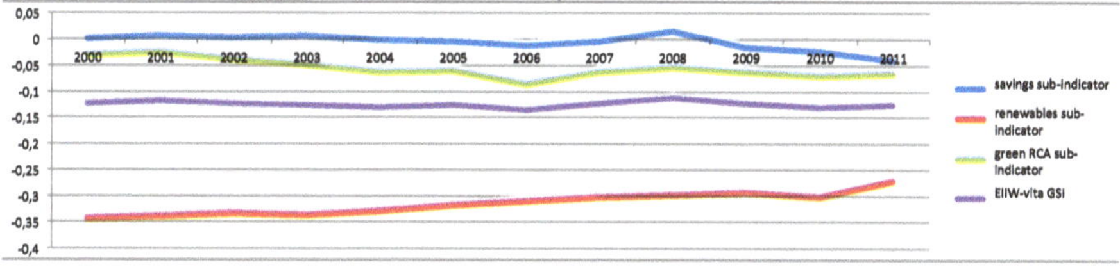
</td>
</tr>
</table>

The United Kingdom has a relatively weak overall performance because of the renewables sub-indicator. Although the sub-indicator has improved over time, it is still negative. 40.24 % of the electricity is produced from natural gas, 30.04 % from coal, and 18.90 % from nuclear sources. The share of renewables in electricity production in the country is only 9.43 %. The UK government tries to promote the production of electricity from renewables. Electricity producers with a capacity of less than 5 MW can sell at a fixed tariff, whereas electricity producers with a capacity of more than 5 MW are obliged to supply their customers with a certain share of electricity from renewable sources (Res Legal 2014). After the economic crisis of 2009, the savings sub-indicator turned to negative and still stays negative in 2011. The country has a good performance as regards water productivity; it is in sixth place worldwide.

Country Profile Greece 2011

GDP (constant 2005 US$) 223,829,580,516.6 **GDP per capita (constant 2005 US$)** 20,122.7	**Adjusted net savings, incl. particulate emission damage (% of GNI)** −5.50 **Savings sub-indicator** −0.1018 (126)
EIIW-vita GSI −0.1144 (121 out of 143) **Extended EIIW-vita GSI incl. water productivity** −0.0919 (125 out of 143)	**Green exports (current US$)** 562,798,840 **RCA sub-indicator** −0.0201 (117)
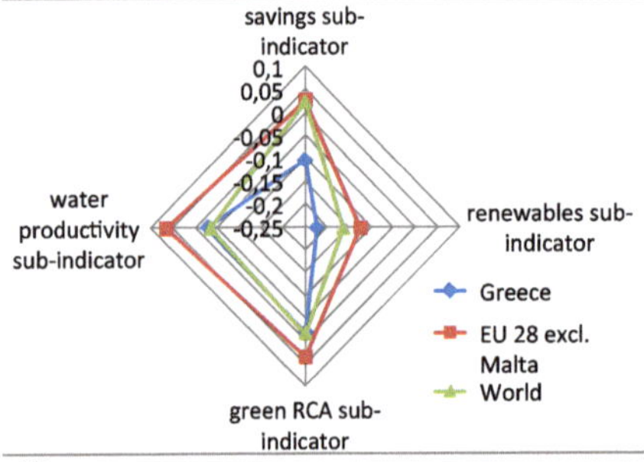	**Share of renewable energy sources in electricity production (% of total)** 13.76 **Renewables sub-indicator** −0.2215 (92) **Water productivity (constant 2005 US$ GDP per m^3 of total freshwater withdrawal)** 23.63 **Water productivity sub-indicator** −0.0242 (54)

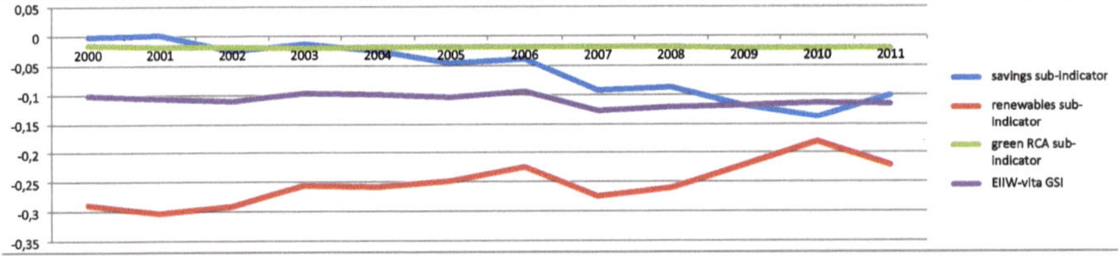

In all sub-indicators, Greece performs worse than the EU average. One can see a negative development in savings sub-indicator since 2000; however, a slight improvement may be seen in 2011. The genuine savings rate in Greece turned from being positive to negative in 2008 and in 2011 is still negative (−5.5 % of GNI). The most important sources for electricity production are coal with 52.50 % and natural gas with 23.56 %. A positive tendency may be seen as regards the use of renewables. The share of renewables in electricity production in Greece is 13.76 %. The producers of electricity from renewables are supported through a feed-in tariff. The use of renewables for heating purposes is promoted by a tax exemption and a subsidy scheme. In transportation, the quota system is used (Res Legal 2014).

Country Profile Italy 2011

<table>
<tr><td>GDP (constant 2005 US$)
1,771,814,582,852.0
GDP per capita (constant 2005 US$)
29,838.9</td><td>Adjusted net savings, incl. particulate emission damage (% of GNI)
6.85
Savings sub-indicator
−0.0014 (74)</td></tr>
<tr><td>EIIW-vita GSI
0.0464 (50 out of 143)
Extended EIIW-vita GSI incl. water productivity
0.0329 (51 out of 143)</td><td>Green exports (current US$)
35,109,674,626
RCA sub-indicator
0.2238 (4)</td></tr>
<tr><td>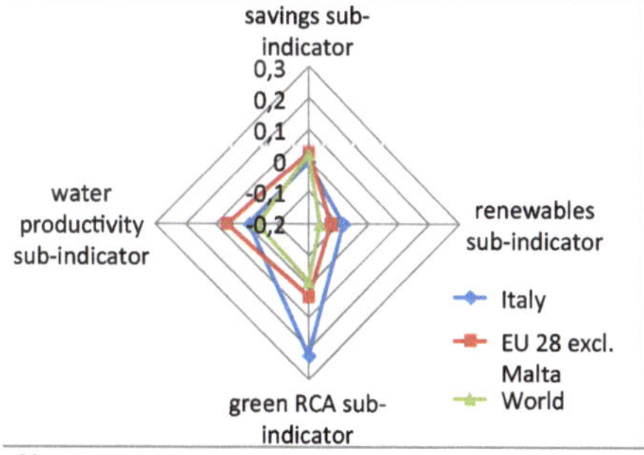
</td><td>Share of renewable energy sources in electricity production (% of total)
27.59
Renewables sub-indicator
−0.0831 (70)
Water productivity (constant 2005 US$ GDP per m³ of total freshwater withdrawal)
38.99
Water productivity sub-indicator
−0.0079 (38)</td></tr>
</table>

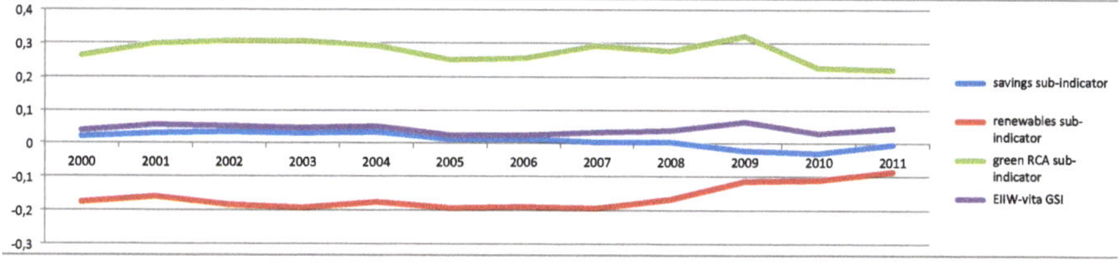

Italy performs above the EU average in "green exports"; it is in fourth place worldwide as regards the RCA sub-indicator. The country is the world's fifth biggest exporter of the "green good." Around 5.5 % of all Italian exports are "green exports." The performance in savings sub-indicator on the other hand is relatively weak. We can see a negative tendency since 2005. In 2009, the indicator turned to be negative. This reflects the influence of the financial crisis on the Italian economy. We can, however, see an improvement in terms of the use of renewables. The share of renewables in electricity production in 2011 is 27.6 %. 48.08 % of electricity is produced from natural gas, 16.68 % from coal, and 6.61 % from oil. It is also worth mentioning that in 2011, Italy did not produce any electricity from nuclear sources at all. The National Energy Strategy (Ministry of Economic Development of Italy 2013) is focused on 4 goals: a reduction of the energy cost gap, the achievement of the environmental targets of Europa 2020, an improvement of security of supply, and the encouragement of sustainable economic growth.

Country Profile Spain 2011

GDP (constant 2005 US$) 1,179,825,331,621.2 **GDP per capita (constant 2005 US$)** 25,240.8	**Adjusted net savings, incl. particulate emission damage (% of GNI)** 9.06 **Savings sub-indicator** 0.0166 (62)
EIIW-vita GSI −0.0351 (78 out of 143) **Extended EIIW-vita GSI incl. water productivity** −0.0290 (77 out of 143)	**Green exports (current US$)** 9,076,231,413 **RCA sub-indicator** −0.0611 (136)
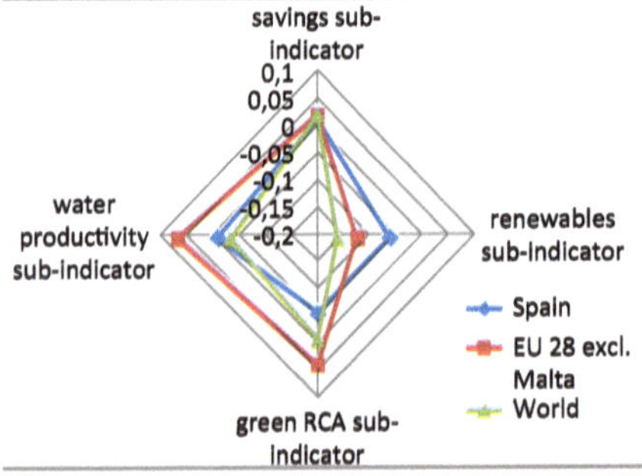	**Share of renewable energy sources in electricity production (% of total)** 29.83 **Renewables sub-indicator** −0.0608 (68) **Water productivity (constant 2005 US$ GDP per m^3 of total freshwater withdrawal)** 36.47 **Water productivity sub-indicator** −0.0105 (43)

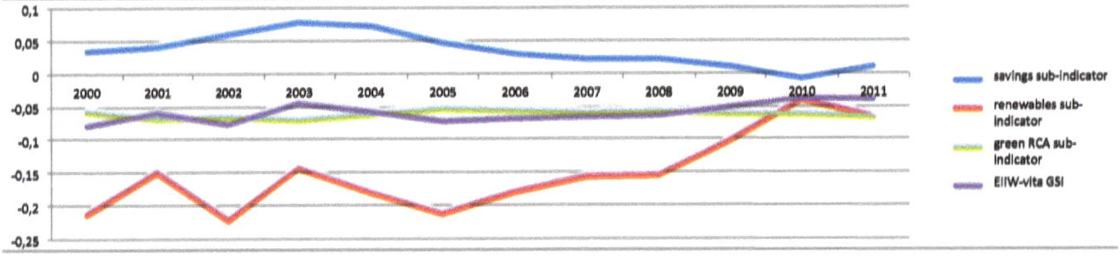

Spain is ranked in 78th place. The saving rate sub-indicator shows a negative tendency since 2003 and turned negative in 2010. However, a slight improvement can be seen in 2011. A great improvement may be seen in the use of renewables: In 2011, Spain produced around 29.83 % of electricity from renewable sources, which makes renewables the most important source for the electricity production in the country. 29.24 % of electricity is generated from natural gas, 19.97 % from nuclear sources, 15.54 % from coal, and 5.11 % from oil. In Spain, renewable energy plants operate under a "Special Regime" which allows priority access and connection to the grid. In addition, the price regulation system promotes the generation of electricity from renewables. Electricity producers have two support options: a guaranteed feed-in tariff or a guaranteed bonus paid on top of the electricity price achieved on the wholesale market (Res Legal 2014).

Country Profile Japan 2011

<table>
<tr><td>GDP (constant 2005 US$)
4,627,423,849,047.6
GDP per capita (constant 2005 US$)
36,203.4</td><td>Adjusted net savings, incl. particulate emission damage (% of GNI)
10.41
Savings sub-indicator
0.0276 (53)</td></tr>
<tr><td>EIIW-vita GSI
0.1647 (15 out of 143)
Extended EIIW-vita GSI incl. water productivity
0.1248 (14 out of 143)</td><td>Green exports (current US$)
73,620,173,305
RCA sub-indicator
0.7033 (2)</td></tr>
<tr><td>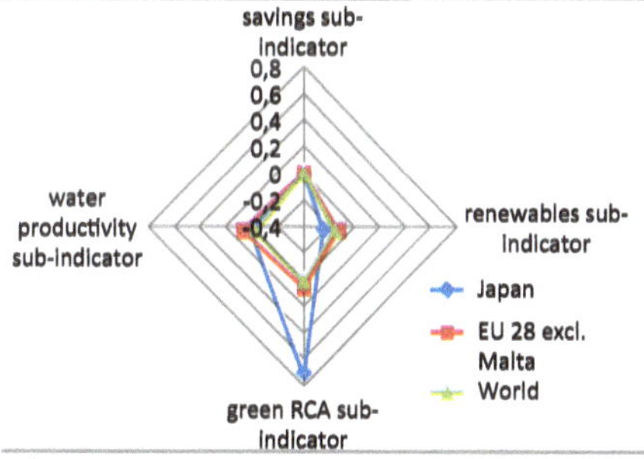
</td><td>Share of renewable energy sources in electricity production (% of total)
12.22
Renewables sub-indicator
−0.2369 (95)
Water productivity (constant 2005 US$ GDP per m^3 of total freshwater withdrawal)
51.33
Water productivity sub-indicator
0.0053 (31)</td></tr>
</table>

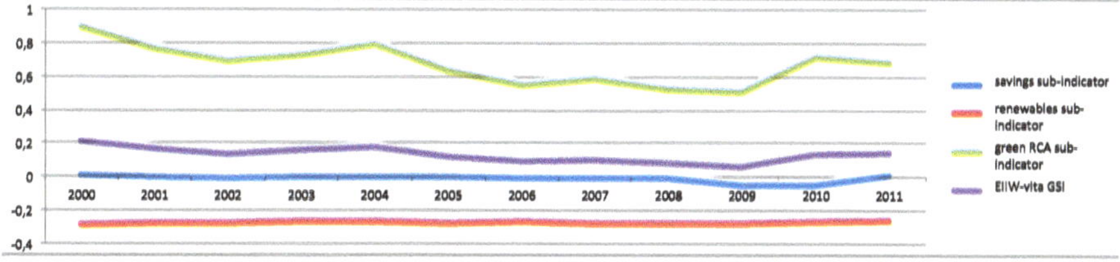

Japan lies in 15th place in the EIIW-vita ranking. The country has a strong performance as regards exports of "green products," being on the second place worldwide. In 2011, "green products" in the value of 73.6 billion US$ were exported from Japan, which make up 8.2 % of the total Japanese exports. The genuine savings rate in Japan fell in the period 2008–2010; however, in 2011 we can see an improvement. As regards the renewables sub-indicator, Japan is below the world and European average: the country produces around 12 % of electricity from renewable sources. In 2011, 35.86 % of electricity is generated from natural gas, 26.96 % from coal, 10.10 % from oil, and 9.76 % from nuclear sources. The Ministry of the Environment of Japan (2014) makes a number of important steps to improve the environment. Among the current initiatives is "Fun to share," which encourages cooperation between companies, organizations, and individuals to share the environmental technologies and knowledge to achieve a low-carbon society.

Country Profile United States 2011

GDP (constant 2005 US$) 13,846,778,425,918.1 **GDP per capita (constant 2005 US$)** 44,440.2	**Adjusted net savings, incl. particulate emission damage (% of GNI)** 0.93 **Savings sub-indicator** −0.0495 (111)
EIIW-vita GSI −0.0694 (97 out of 143) **Extended EIIW-vita GSI incl. water productivity** −0.0570 (97 out of 143)	**Green exports (current US$)** 71,959,498,495 **RCA sub-indicator** 0.0781 (5)
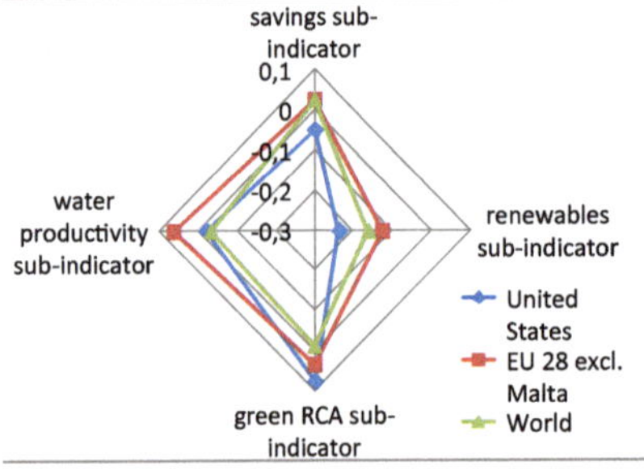	**Share of renewable energy sources in electricity production (% of total)** 12.23 **Renewables sub-indicator** −0.2368 (94) **Water productivity (constant 2005 US$ GDP per m³ of total freshwater withdrawal)** 27.65 **Water productivity sub-indicator** −0.0200 (48)

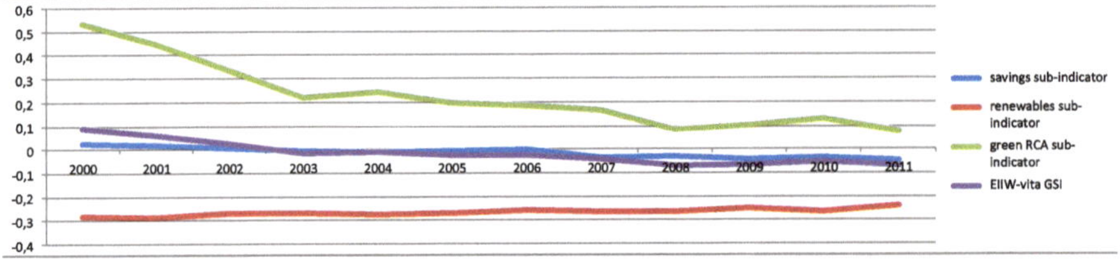

The United States is on 97th place in the ranking. The "green competitiveness" sub-indicator is above both the world and European average, but shows a negative tendency. As the indicator is relative, negative dynamics in the USA can be explained by the better performance of developing economies like China. In absolute terms, the USA was the fourth biggest producer of "green products" worldwide after Germany, China, and Japan. The share of "green exports" in total US exports is around 3.4 %. In the use of renewables, the US is on the 94th place. 43.35 % of the electricity in the country in 2011 is produced from coal, 24.16 % from natural gas, 18.98 % from nuclear sources, and 12.23 % from renewable sources. As regards the true savings rate sub-indicator, the USA has negative values in the period 2003–2010, except for the year 2006, and hence it can be concluded that the USA faces problems with sustainable growth. Important environmental initiatives of the USA can be found on the website of the EPA (2014).

References

Brazilian Electricity Regulatory Agency (2013) The Brazilian electricity regulatory agency. http://www.aneel.gov.br. Accessed 21 Jan 2015

Department of Energy of South Africa (2013) Renewable energy resources in South Africa. http://www.irena.org/DocumentDownloads/events/2013/July/Africa%20CEC%20session%203_RSA%20Department%20of%20Energy_Modise_220613.pdf. Accessed 21 Jan 2015

EPA (2014) United States environmental protection agency. http://www.epa.gov. Accessed 21 Jan 2015

European Environment Agency (2012) Towards efficient use of water resources in Europe. EEA Report 1/2012. doi:10.2800/95096

EY (2013) Mapping India's renewable energy growth potential: status and outlook 2013. http://www.ey.com/Publication/vwLUAssets/Mapping_Indias_Renewable_Energy_growth_potential$FILE/EY-Mapping-Indias-Renewable-Energy-growth-potential.pdf. Accessed 21 Jan 2015

Fifth BRICS Summit (2014) BRICS. http://www.brics5.co.za. Accessed 21 Jan 2015

Government of South Africa (2011) Integrated resource electricity plan 2010–2030

Government of Indonesia (2006) Presidential regulation no. 5/2006. Jakarta

Hanemann WM (2006) The economic conception of water. In: Rogers P, Ramon Llamas M, Martinez-Cortina L (eds) Water crisis: myth or reality? Taylor & Francis plc, London

IEA (2014) Policies and measure, renewable energy. http://www.iea.org/policiesandmeasures/renewableenergy/?country=France. Accessed 21 Jan 2015

IMF (2014) Data and statistics. http://www.imf.org. Accessed 21 Jan 2015

Ministry of Economic Development of Italy (2013) Italy's National Energy Strategy: for a more competitive and sustainable energy. http://www.encharter.org/fileadmin/user_upload/Energy_policies_and_legislation/Italy_2013_National_Energy_Strategy_ENG.pdf. Accessed 21 Jan 2015

Ministry of Economy and Energy of Germany (2014) Information portal renewable energy. http://www.erneuerbare-energien.de. Accessed 21 Jan 2015

Ministry of Energy and Natural Resources (2006) Blueprint Pengelolaan Energi Nasional 2006–2025. Jakarta

Ministry of the Environment of Japan (2014) Ministry of the environment of Japan. http://www.env.go.jp/. Accessed 21 Jan 2015

RES LEGAL (2014) Legal source on renewable energy. http://www.res-legal.eu. Accessed 21 Jan 2015

State Council Information Office of the People's Republic of China (2012) White Paper: China's Energy Policy (2012). China.org.cn

United Nations Commodity Trade Statistics database, Department of Economic and Social Affairs/Statistics Division. http://comtrade.un.org/db/. Accessed 21 Jan 2015

World Bank, World Development Indicators Database. http://data.worldbank.org. Accessed 21 Jan 2015

7 Why the Global Sustainability Perspective Is Important for Asian Countries

For the last decade, Asian countries have been enjoying a positive and steady economic growth. It is projected that the economy of Asia will grow steadily at 5.4 % in 2014 and 5.5 % in 2015 (IMF 2014). Despite some disruptions, such as the global economic crisis, the Asian economy still grew positively due to its strong domestic demand (IMF 2014). However, one question remains, "Is Asian economic growth sustainable?" Here one may use the EIIW-vita Global Sustainability Indicator to monitor current progress with respect to the sustainability of Asian economic growth.

Sustainable development is not a new economic model for Asian countries. Many Asian countries have adopted the concept and several initiatives have already been introduced both individually and multilaterally. China introduced the Comprehensive Working Plan for Energy Conservation and Emission Reduction in August 2011. Two years before that, the Singaporean government introduced the Sustainable Singapore Blueprint. The Asian Development Bank (ADB) also offered support to all Asian countries to implement a sustainable development framework through three strategic agendas in the ADB Strategy 2020, namely, inclusive economic growth, environmentally sustainable growth, and regional integration (ADB 2008).

A recent OECD report emphasizes the importance for Asian countries (particularly Southeast Asia) to act now (OECD 2014). OECD (2014) argues that Southeast Asian countries will miss three golden opportunities if they delay their green growth initiatives: (1) to sustain the region's natural wealth, (2) to lock in clean and resilient infrastructure, and (3) to become a hub for green investment. Besides those three factors, sustainable development is very important for Asia due to at least two further reasons. Firstly, Asia supports nearly 60 % of the world's population and contributed approximately 33 % of total world output in 2013. Secondly, during the last decade, we have witnessed the increasing frequency and scale of impacts of climate-related disasters, such as floods in Bangkok (Thailand) and Jakarta (Indonesia), several storms in the Philippines, and many more. This chapter presents the rationale behind the suggestion that Asian countries should consider

P.J.J. Welfens et al., *Towards Global Sustainability*,
DOI 10.1007/978-3-319-18666-5_7

the Global Sustainability Indicator to monitor their progress on the implementation of a sustainable development framework. Moreover, this chapter also presents selected "green growth" initiatives (particularly in renewable energy) in selected Asian countries.

Demographic and Economic Pressure on the Environment in Asia

The role of Asian economies in terms of the world's GDP is becoming increasingly significant over time. Figure 7.1 shows the increasing trend of Asia's contribution to global GDP, from only 15 % in 1971 to 33 % in 2013. As regards GDP growth, Asian economies grew at a faster rate than the world's economy except during the Asian Financial Crisis of 1998. By comparing two crises, the 1998 Asian Financial Crisis and the 2008 Global Financial Crisis, the Asian economy seemed to be more resilient to the crisis. Asian economic growth dropped significantly from 3.8 % in 1997 to −0.01 % in 1998. Moreover, the contribution of Asia's GDP to the world's GDP also fell from 27 % in 1997 to 25 % in 1998. Interestingly, even though the global financial crisis had a negative impact on Asian economies, the impact was not as severe as that which was experienced by non-Asian countries (such as European countries and the USA). The Asian economy still had positive growth, whereas the world's GDP shrank to minus 2 % in 2009 relative to the previous year. The contribution of Asia to global GDP increased from 27 % in 2008 to 29 % in 2009.

Asia is not only important in terms of its role in the world's GDP but also in terms of population. Based on the UNCTAD database, Asia is home to nearly 60 % of the world's population. Prior to 1999, population growth in the Asia region was always higher than global population growth. Then, it continuously dropped until it fell below the rate of global population growth in 1999 (see Fig. 7.2). Since then,

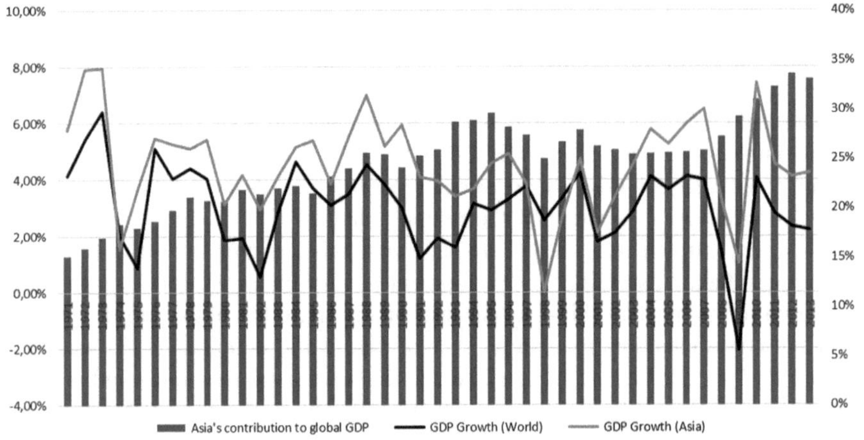

Fig. 7.1 The role of Asia in the global economy. *Data Source*: UNCTAD Database, accessed January 2015

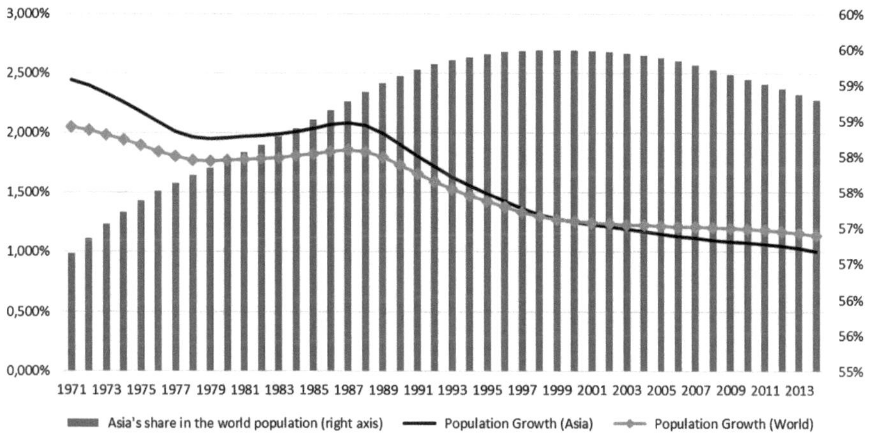

Fig. 7.2 The population of Asia and its growth. *Data Source*: UNCTAD Database, accessed January 2015

Asia's population growth rate has been lower than the world's population growth rate. The United Nations has predicted that the total population in Asia will increase from 4.2 billion people in 2013 to roughly 5.1 billion people in 2050 (UN 2012). It is also predicted that the population of India will surpass China's population, making India the most populous country in the world.

The increasing regional output in Asia will demand more input, particularly of raw materials and energy which are extracted directly from the earth. Moreover, it will also result in more pollutants which will be released during the production process. Together, the increasing regional output can be translated into more pressure on the environment. Indeed, it will also depend on the structure of the economy and energy intensity of Asian countries.

Natural capital contributes significantly to the total wealth of Asian countries. The evidence is not only found in low-income Asian countries (such as Lao PDR) but also in Asian middle-income (such as Indonesia) and high-income countries (such as Brunei Darussalam). The production of oil and natural gas accounts for almost 75 % of the total wealth of Brunei Darussalam. Oil and natural gas are also important for Indonesia, Thailand, and Vietnam, accounting for as much as 23 %, 8 %, and 24 % of total natural capital, respectively. The shares of crop production and timber are significantly high in Lao PDR. Together they account for as much as 38 % of the total wealth of Lao PDR (see Fig. 7.3).

The main challenges faced by Asian countries are not only the high dependency on natural capital but also the depletion of natural capital.[1] OECD (2014) reported

[1] OECD (2014) defines natural capital depletion as the sum of net forest depletion, energy depletion, mineral depletion, and particulate emission damage. Net forest depletion is computed as the product of unit resource rents and the excess of round wood harvest over natural growth. Energy depletion is calculated as the ratio of the value of the stock of energy resources, which

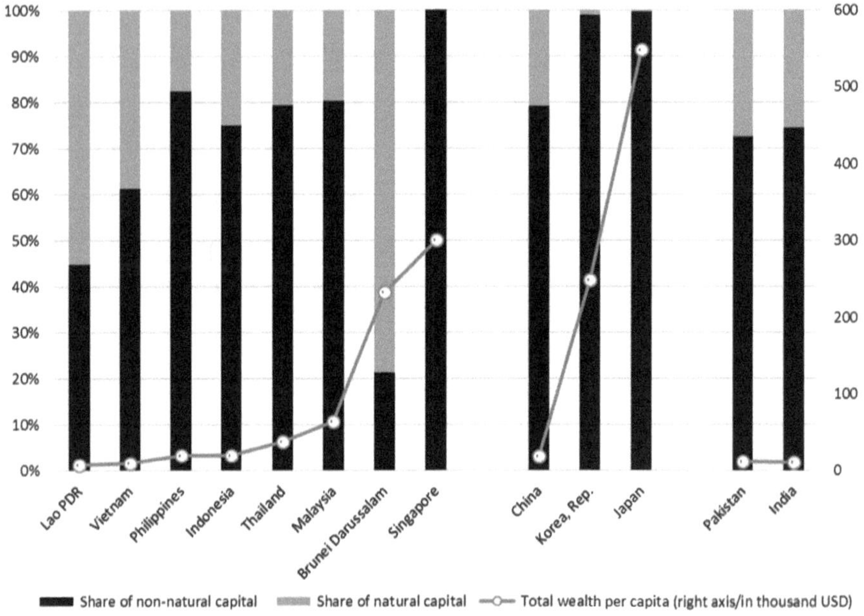

Fig. 7.3 The contribution of natural capital and nonnatural capital to the total wealth in selected Asian countries, 2005. *Note*: Total wealth per capita is in thousands of 2005 USD; nonnatural capital includes net foreign assets, intangible capital, and produced capital; natural capital includes crop, pasture land, timber, non-timber forest, protected areas, oil, natural gas, coal, and minerals. *Data Source*: World Bank, The Wealth of Nation data set, accessed January 2015

that in some Asian countries, natural capital is being depleted at an increasing rate. The most worrying case is Brunei Darussalam. The natural capital stock of Brunei Darussalam was depleted by nearly 40 % of gross national income (GNI) on average every year during the period 1999–2012. This was substantially larger than the depletion of natural capital stock in other Asian countries such as Malaysia (9.1 % of GNI), Indonesia (7.7 % of GNI), and Vietnam (8.8 % of GNI). Natural capital depletion in China and India was quite moderate, at 3.9 % of GNI and 3.7 % of GNI, respectively. In more detailed analysis, the type of natural capital which was rapidly depleted in most Asian countries (including Brunei Darussalam, Indonesia, Malaysia, Thailand, Vietnam, China, and India) was energy.

As previously mentioned, the pressure on the environment due to economic activities will also depend on energy intensity. In general, there have been improvements in terms of energy intensity in the Asian countries (see Fig. 7.4). Myanmar is the country with the most progress in energy intensity improvement.

cover coal, crude oil, and natural gas, to the remaining reserve lifetime. Mineral depletion is computed as the ratio of the value of the stock of mineral resources to the remaining reserve lifetime. Types of energy that are covered in this variable are tin, gold, lead, zinc, iron, copper, nickel, silver, bauxite, and phosphate. The remaining reserve lifetime is assumed equal to 25 years.

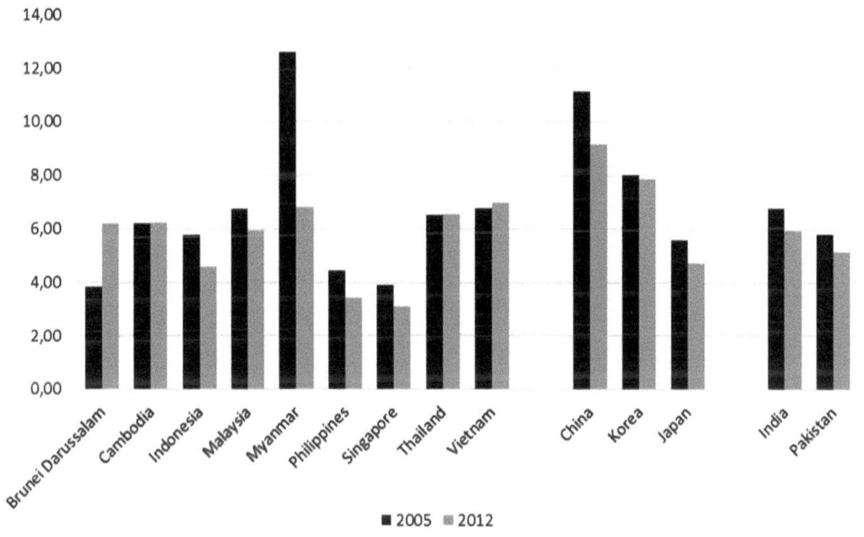

Fig. 7.4 Energy intensity of selected Asian countries. *Note*: Energy intensity is expressed as the ratio of total primary energy supply (in petajoules) per GDP in USD 2005 purchasing power parity. *Data Source*: International Energy Agency, CO_2 Emissions from Fuel Combustion Statistics 2014

Its energy intensity dropped significantly from 12.6 in 2005 down to 6.8 in 2012. Myint and Aung (2013) predicted that Myanmar's energy intensity will drop continuously at an average rate of 1.3 % per year over the period 2010–2035.[2] On the other hand, Brunei Darussalam's energy intensity increased substantially from 3.8 in 2005 up to 6.2 in 2012. Even though the absolute value of the total primary energy supply of Brunei Darussalam is the smallest relative to other selected Asian countries, more actions from the Government of Brunei Darussalam are necessary in order to bring the country back onto a sustainable development path.

As regards energy, Asia has another important issue to come to grips with. Asia has to reduce its dependency on nonrenewable energy. Based on the EIA database, the share of nonrenewable energy in the Asian energy mix increased from 75 % in 2002 to 83 % in 2012. This does not necessarily mean that Asia consumed less renewable energy in 2012 relative to 2002. In absolute values, the total final consumption of renewable energy in Asia increased by 31 % in 2012 relative to 2002. However, this was far behind the growth in the final consumption of nonrenewable energy which increased by more than twofold (110 %) in 2012 relative to 2002. Figure 7.5 suggests that the share of natural gas in the Asian energy mix increased slightly, by 1 % point, from 7 to 8 % in 2012 relative to 2000. The

[2] The results are based on a business-as-usual scenario in a CGE analysis which is conducted by Myint and Aung (2013). In this scenario, they assumed that Myanmar's GDP will grow at an average annual rate of around 7 % during the period 2010–2035. It was also assumed that population will increase by 1 % per year during the same period.

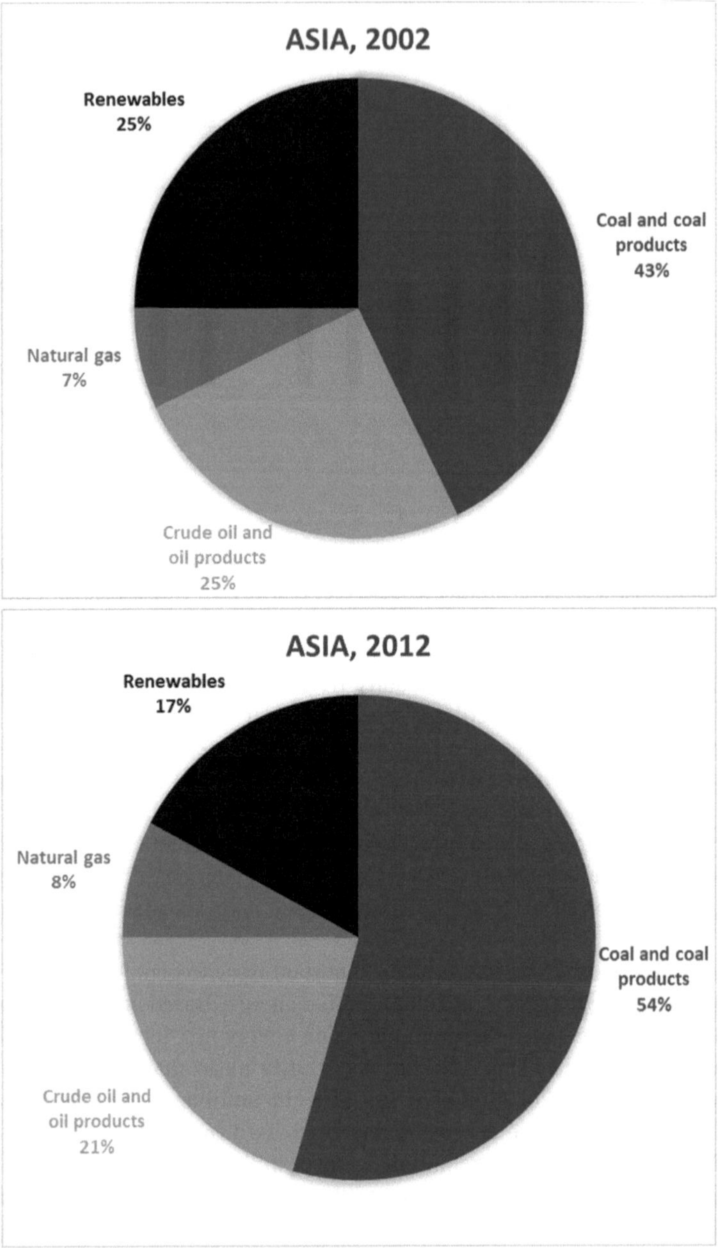

Fig. 7.5 Energy mix in Asia, 2002 and 2012. *Note*: Energy mix is based on total primary energy supply; renewables cover nuclear, hydropower, geothermal, solar/wind/other, biofuels, and waste (including industrial waste, municipal waste, charcoal, biogasoline, biodiesel, other liquid biofuels, solid biofuels, and biogases). *Data Source*: International Energy Agency, Energy Balance of Non-OECD countries, 2014

contribution of coal and coal products in the Asian energy mix also increased from 43 % in 2000 to 54 % in 2012. On the other hand, the share of crude oil and oil products in the Asian energy mix dropped from 25 % in 2000 to 21 % in 2012.

The share of renewable energy in the Asian energy mix looks really promising. However, one should note that biofuels in Fig. 7.5 cover all types of biofuels including solid biofuels, such as firewood, wood chips, bark, sawdust, shavings, and other solid biomass which are not environmentally friendly. Pitt (1985) found that the utilization of wood fuel in Java (Indonesia) caused serious damage to the environment due to deforestation. Thus, a second version of the Asian energy mix is presented in Fig. 7.6, which excludes waste, biogases, and solid biofuels. Figure 7.6 shows that renewable energy makes the smallest contribution to the Asian energy mix relative to other types of energy. Moreover, the transition process from nonrenewable energy to renewable energy is proceeding at a slow pace. The share of renewables increased slightly from 4.5 % in 2002 to 5.1 % in 2012.

Although the share of renewable energy in the Asian energy mix in 2012 was lower than 2002, in terms of absolute value, the total supply of all types of renewable energy increased over time. Figure 7.7 shows the increasing trend of all types of renewable energy, namely, nuclear, hydropower, geothermal, solar/wind/other, and biofuels. The contribution of hydropower is the largest compared to other types of renewable energy. The share of hydropower is always above 44 % of the total supply of renewables. The second most important renewable energy in Asia is nuclear power. It accounted for as much as 22 % of the total supply of renewable energy in Asia in 2012. Nuclear power is quite essential as a source of energy in East Asia and South Asia. It should be noted, however, that there is no operational nuclear power plant in Southeast Asia up to 2012. Five members of the Association of Southeast Asian Nations (ASEAN) have announced plans to build nuclear power plants, namely, Indonesia, Malaysia, Thailand, Vietnam, and the Philippines (Nugroho 2011).

Vietnam is the most aggressive country in terms of the development of nuclear power plants in Southeast Asia. They have announced plans to build ten nuclear power plants by 2030 with a total capacity of 10,700 MW (Daiss 2012). Currently, Vietnam has signed agreements with Russia and Japan to build two power plants. Russia has agreed to finance the first of Vietnam's nuclear power plants at Phuoc Dinh in Ninh Thuan province. It had been planned for construction to begin in 2014 and the plant to be ready to operate in 2020. Unfortunately, construction has been delayed until 2017. Japan has also committed to help Vietnam build another power plant at Vinh Hai in Ninh Thuan province. This project has also been delayed several years from its original plans (December 2015). The full story can be found on the website of the World Nuclear Association (2014).

The development of renewable energy in Asia is country specific (see Appendix A). Although Brunei Darussalam focuses on the development of solar panels, among ASEAN member states, the development of renewable energy in Brunei Darussalam is left behind. There is only one electricity power plant that uses renewable energy, namely, the 1.2 MW Tenaga Suria Brunei solar plant in Seria [see the details on the website of the Oxford Business Group (2013)]. This project,

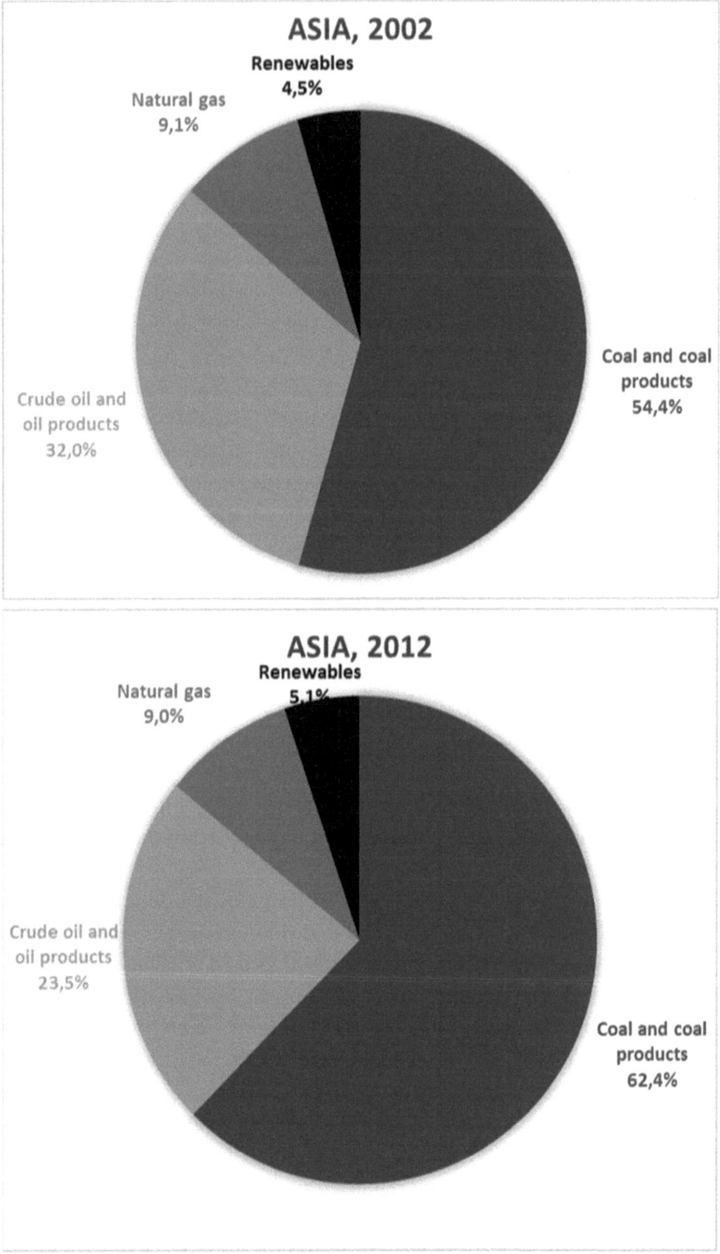

Fig. 7.6 Energy mix in Asia after excluding waste, biogases, and solid biofuels, 2002 and 2012. *Note*: Energy mix is based on total primary energy supply; renewables cover nuclear, hydropower, geothermal, solar/wind/other, biofuels, and waste (including biogasoline, biodiesel, and other liquid biofuels). *Data Source*: International Energy Agency, Energy Balance of Non-OECD countries, 2014

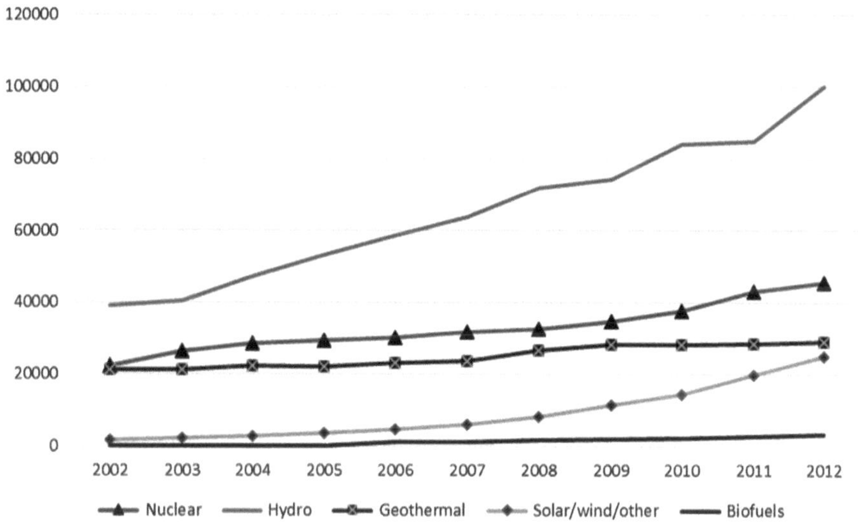

Fig. 7.7 Total supply of renewable energy in Asia by types of sources, 2002–2012. *Note*: Biofuels cover biogasoline, biodiesel, and other liquid biofuels. *Data Source*: International Energy Agency, Energy Balance of Non-OECD countries, 2014

started in May 2011, represents only 1 % of total installed capacity. The private sector is also involved in the development of renewable energy in Brunei Darussalam by initiating several small projects (IRENA 2013a), such as a 5 kW solar photovoltaic roof off-grid system (1 project), 29.9 kW solar photovoltaic roof grid-connected pilot projects (7 projects), and 3.23 kW solar photovoltaic systems at offshore oil/gas platforms (20 projects). The Government of Brunei Darussalam has set a relatively modest target on the development of renewable energy. Electricity generation from renewable energy is targeted to reach 10 % of total installed capacity in 2035. The lack of motivation of Brunei Darussalam to substitute nonrenewable energy with renewable energy is understandable, since they have abundant oil and natural gas reserves. Similar to Brunei Darussalam, Singapore also has made small progress on the development of renewable energy. Up to 2012, Singapore has only focused on the development of solar energy, the main reason being due to the size of the country which is very small. Singapore has no significant energy endowment and depends largely on imported fossil fuels (Finenko 2014).

Cambodia, Myanmar, and Vietnam (CMV) focus largely on the development of hydropower. Cambodia and Vietnam have also been developing solar and wind energy. However, their contribution relative to the total supply of renewable energy is less than 1 %. Among new ASEAN member states, Vietnam is the most optimistic country and has been preparing for the construction of the first nuclear power plant in Southeast Asia. In 2014, the Government of Vietnam sent 344 Vietnamese to Russia to learn about nuclear power plants as part of the agreement with the Government of Russia. The details can be found on the website of the World Nuclear Association (2014).

Hydropower also makes up a large share of the supply of renewable energy in Malaysia. It accounted for as much as 87 % of the total supply of renewable energy in Malaysia in 2012. Since 2006, Malaysia has been developing biofuels as an energy alternative, particularly biodiesel based on palm oil. Together with Indonesia, Malaysia is a world leader in the palm oil industry. Biofuels (in the form of biodiesel) also played a significant role in the supply of renewable energy in Indonesia. Indeed, the contribution of biofuels is not the highest one among all types of renewable energy. However, the development has been quite progressive in the last 6 years. Total supply of biofuels increased substantially from only 4.4 kt of oil equivalent (ktoe) in 2006 up to 523.9 ktoe in 2012. The most important source of renewable energy in Indonesia is geothermal. It accounted for about 90 % of the total supply of renewable energy in 2012. It is expected that the development of geothermal energy will be progressive in the future in light of the introduction of new regulations on geothermal energy production on August 26, 2014. There are several benefits from the new regulations of geothermal energy production. Firstly, exploitation activity is possible to be conducted in protected forests. Approximately 21 % of geothermal sources in Indonesia are located in protected forests. Secondly, all the permits for a geothermal extraction process will be issued by the central government, this is aimed at reducing the bureaucracy and time required to start extraction activity. Lastly, the Government will assign several government-owned corporations to develop geothermal activities in locations which are not attractive for other investors.

The Philippines also has big concerns with respect to the development of geothermal power. The share of geothermal in the total supply of renewables was 88 % in 2012. Even though the total reserve of geothermal in the Philippines is much lower than is the case with Indonesia, the Philippines has succeeded in developing geothermal since 1977 (Birsic 1980). The first geothermal power plant in the Philippines was the 3000 kW Leyte Geothermal Pilot Plant. In 1979 the Philippines built their second geothermal power plant, the Tiwi Geothermal Electric Plant, which has an installed capacity of as much as 55,000 kW. Based on the International Geothermal Association, the Philippines ranked second in the world after the USA in terms of the development of geothermal power plants (Dickson and Fanelli 2004).

Unlike Southeast Asia, countries in East Asia (i.e., China, Japan, and Korea) depend largely on the nuclear power plant. Before the tragedy in Fukushima, nuclear power contributed more than 87 % of the total supply of renewable energy in Japan. Then, the Government of Japan shut down most of its nuclear power plants due to security reasons. Consequently, the share of nuclear power in the total supply of renewable energy in Japan dropped significantly from 70 % in 2011 to 15.8 % in 2012. Korea also has a large dependency on nuclear power since the share of nuclear to the total supply of renewable energy was about 97 % in 2012. The development of other types of renewable energy was quite modest in Korea. There was moderate progress in the development of solar power and biofuels but it was substantially lower relative to nuclear-based energy.

Even though nuclear power is an important source of renewable energy for China, it nevertheless lags far behind hydropower. The share of hydropower to the total supply of renewable energy in China accounted for as much as 59 % in 2012. The second most important source is nuclear which accounted for about 20 % of total supply of renewable energy in 2012. China has also been developing solar and wind energy with a contribution from these sources of as much as 17 % of the total supply of renewable energy. There was a progressive trend in the development of solar and wind energy. In 2002, the total supply of solar and wind energy was only 1510 ktoe. Since then, it has increased significantly, reaching 21,683 ktoe in 2012.

Nuclear energy is also the second most important renewable energy source for South Asian countries. The most important source of renewable energy for Pakistan and India is hydropower which accounted for as much as 68 % and 53 %, respectively, of the total supply of renewable energy in these countries. The shares of nuclear energy in Pakistan and India were 32 % and 43 % of the total supply of renewable energy, respectively. Up to 2012, there was no development of solar and wind energy and biofuels in Pakistan. Similarly, the development of solar and wind energy and biofuels has been really slow in India.

The pressure on the environment is not only coming from economic activities but also from the increasing population. The increasing number of people can be translated into an increasing demand for space (land), food, and water. Moreover, it will also increase the vulnerability to the impact of natural disasters. This issue is further discussed in the next subchapter. In terms of the impact of population growth on food and water, one should focus on the availability of water. Water is essential for all humankind, not only for basic needs (such as drinking water) but also as an input for agriculture and industry.

Table 7.1 shows the coverage of drinking water for selected Asian countries. Generally, access to improved water had progressed in 2012 relative to 1990. However, the proportion of people with access to improved water varies across countries. Twenty-nine percent of Cambodians still have no access to improved water in 2012. Indeed, this state of affairs is much better relative to the situation in 1990, when 88 % of Cambodians lacked of access to improved water. Myanmar and Indonesia also face the same problem with approximately 14 % of their citizens still having no access to improved water. An impressive performance with respect to access to improved water can be found in Vietnam. The number of Vietnamese who lack access to improved water reduced significantly from 38 % in 1990 to only 5 % in 2012. A similar pattern is also found in the case of both China and India. The coverage of improved drinking water in China increased significantly from 67 % in 1990 to 92 % in 2012.

Table 7.1 Drinking water coverage estimates, 1990 and 2012 (in percentage)

Country	Year	Piped onto premises (%)	Other improved sources (%)	Other unimproved (%)	Surface water (%)
Cambodia	1990	2	20	42	36
	2012	18	53	15	14
Indonesia	1990	9	61	24	6
	2012	21	64	13	2
Malaysia[a]	1990	86	8	6	0
	2012	99	1	0	0
Myanmar	1990	5	51	17	27
	2012	8	78	11	3
Philippines	1990	24	60	14	2
	2012	43	49	7	1
Singapore	1990	100	0	0	0
	2012	100	0	0	0
Thailand	1990	29	57	12	2
	2012	48	48	4	0
Vietnam	1990	9	53	22	16
	2012	26	69	4	1
China	1990	33	34	26	7
	2012	71	21	7	1
Korea[a]	1990	96	1	3	0
	2012	99	1	0	0
Japan	1990	94	6	0	0
	2012	98	2	0	0
Pakistan	1990	23	62	7	8
	2012	36	55	6	3
India	1990	17	53	27	3
	2012	26	67	6	1

[a]The data is available only for urban area

Data Source: WHO/UNICEF, Joint Monitoring Programme for Water Supply and Sanitation, 2014

Climate Change, Extreme Weather Events, and the Increasing Vulnerability of Asia

The relationship between climate change and an increasing trend of climate-related disaster events has been greatly discussed by many scientists. Karl et al. (1997), Mitchell et al. (2006), Huppert and Sparks (2006), and Anderson (2006) have noted the possible effect of climate change on the increasing frequency of extreme weather events. Those studies are also supported by a World Bank report titled "Turn Down the Heat: Climate extremes, Regional Impacts, and the Case for Resilience" (2013). In this report, the World Bank identified several extreme weather events in a regional-based analysis which are strongly correlated to climate change. Southeast Asia is expected to experience heat extremes, a sea-level rise,

Table 7.2 Possible sector based and thematic impacts due to climate-change-related extreme weather events

Southeast Asia	South Asia
River deltas are expected to be impacted by projected sea-level rise and increases in tropical cyclone intensity	Crop yields are vulnerable to a host of climate-related factors in the region
Fisheries would be affected	Total crop production and per capita calorie availability is projected to decrease significantly
Aquaculture farms may be affected by several climate change stressors	Water resources are already at risk in the densely populated countries of South Asia
Coral reef loss and degradation would have severe impacts for marine fisheries and tourism	Deltaic regions and coastal cities are particularly exposed to compounding climate risks
Agricultural production, particularly for rice in the Mekong Delta, is vulnerable to sea-level rise	Energy security is expected to come under increasing pressure from climate-related impacts to water resources
Coastal cities concentrate increasingly large populations and assets exposed to climate change risks	

Source: The World Bank, Turn Down the Heat: Climate Extremes, Regional Impacts, and the Case for Resilience, 2013

tropical cyclones, and saltwater intrusion. Moreover, the World Bank (2013) also predicted several extreme weather events which might happen in South Asia, namely, heat extremes, precipitation, monsoons, drought, glacial loss, sea-level rise, and snow cover reductions. Together these will cause several negative impacts on each region (see Table 7.2).

Asian countries should increase their awareness of the climate change issue due to its potentially severe impacts on both the people and the economy. A study by Dellink et al. (2014) reported that the effect of climate change impacts on annual global GDP is increasing over time. It is projected that by 2060 the total loss in GDP due to climate change will be between 0.7 and 2.5 %. Moreover, the study also predicted that large negative consequences of climate change will be experienced by South and Southeast Asian countries. Based on the EM-DAT database, Asia accounted for 33 % of total recorded disasters in the world and about 75 % of total affected individuals globally in 2014. This implies that Asian populations are particularly vulnerable to natural disasters. Several factors could play a role in causing this vulnerability in Asia. One of the factors is the uneven distribution in terms of population dynamics in Asian countries. Most Asian people are willing to reside in urban areas due to economic motives, without considering the increasing environmental burden in that area. The populations in Asia seem to be centralized in big cities such as Jakarta (Indonesia), Bangkok (Thailand), or Manila (Philippines).

Most natural disasters do not happen in Asia; however, Asia has the largest share of total affected individuals due to natural disasters relative to other regions. Table 7.3 shows the ratio of natural disaster events and their impact in Asia relative

Table 7.3 Ratio of natural disaster events and their impact in Asia relative to the world (in percentage)

Year	Occurrence (%)	Deaths (%)	Affected (%)	Injured (%)	Homeless (%)	Total_affected (%)	Total_damage (%)
1975–1984	27	54	67	47	38	67	19
1985–1994	27	52	79	50	58	79	31
1995–2004	23	38	84	71	67	83	43
2005–2014	24	42	73	57	64	73	40
2011	21	63	69	63	87	70	63
2012	25	33	61	77	24	61	16
2013	27	49	50	65	64	50	34
2014	33	40	75	60	84	75	71

Note: Natural disaster events cover biological disasters (epidemic, insect infestation), climatological disasters (drought, wildfire), geophysical disasters (earthquake, mass movement, volcanic activity), hydrological disasters (flood, landslide), and meteorological disasters (extreme temperature, storm)
Data Source: The EM-DAT Database, accessed January 2015

Table 7.4 The frequency of natural disasters based on types of disaster in Asia (in percentage)

Year	Biological (%)	Climatological (%)	Geophysical (%)	Hydrological (%)	Meteorological (%)
1977–1984	7	11	15	34	33
1985–1994	7	3	15	38	37
1995–2004	11	4	13	40	32
2005–2014	3	3	13	50	31
2011	4.3	2.2	15.1	51.6	26.8
2012	2.6	1.3	21.1	44.7	30.3
2013	0.0	2.2	13.2	49.4	35.2
2014	0.0	1.1	20.2	37.2	41.5

Data Source: The EM-DAT Database, accessed January 2015

to the world. It presents both an average value for 10-year intervals from 1975 to 2014 and the statistics for the last 4 years individually. Interestingly, natural disasters in Asia do not occur as often as natural disasters in other regions, such as North America. However, the ratio of total affected individuals in Asia is extremely large, with values ranging from 50 to 87 % relative to total affected individuals in the world. This implies that large numbers of Asian people are living in vulnerable areas. Moreover, it can also be understood in terms of the lack of readiness of Asian populations to face natural disasters. If we ruled out developed Asian countries, most other Asian countries are still facing the absence of both good supportive infrastructures and better crisis management systems.

The types of natural disasters which happen most often in Asia are hydrological and meteorological disasters. Table 7.4 suggests that more than 65 % of natural disasters in Asia are hydrological and meteorological related. This percentage is even higher in the last 4 years, with those two forms of event making up 74 % of all natural disasters in Asia. Examples of these types of natural disaster are flooding, landslides, extreme temperatures, and storms. These types of natural disasters are some examples of climate-change-related disasters as mentioned in the report of the World Bank (2013). Thus, one may conclude that Asia shoulders some of the burden related to the negative impact of climate change.

Hydrological disasters are also the type of natural disasters which have the largest scale of impact based on number of total affected individuals. Based on the EM-DAT database, more than half of the victims of natural disasters in Asia are individuals affected by a hydrological natural disaster (see Table 7.5). One exemption to this rule of thumb is 2014 during which a lot of Asian countries experienced flooding, such as China, India, Sri Lanka, and Pakistan. It is estimated that more than 27 million individuals were affected due to floods and landslides in Asia.

The economic impact of natural disasters is also significant. It is estimated that the economic cost of natural disasters in 2014 amounts to US$58.9 billion. The

Table 7.5 Total affected individuals due to natural disasters based on types of disaster in Asia (in percentage of total victims of disasters)

Year	Biological (%)	Climatological (%)	Geophysical (%)	Hydrological (%)	Meteorological (%)
1977–1984	1	26	1	57	15
1985–1994	0	15	2	67	17
1995–2004	0	23	2	63	12
2005–2014	0	11	4	60	25
2011	0.0	16.8	0.1	81.1	1.9
2012	0.0	1.2	0.9	76.8	21.0
2013	0.0	7.4	1.6	74.0	17.0
2014	0.0	10.3	7.7	28.0	54.1

Data Source: The EM-DAT Database, accessed January 2015

Table 7.6 Total economic costs due to natural disasters based on types of disaster in Asia (in thousand USD)

Year	Climatological	Geophysical	Hydrological	Meteorological
1977–1984	202,000	2,663,000	4,584,913	2,747,519
1985–1994	14,977,487	10,972,100	56,885,454	36,243,211
1995–2004	9,599,487	153,420,492	136,489,641	67,265,739
2005–2014	9,274,000	335,770,303	140,586,908	103,333,803
2011	142,000	210,346,400	15,480,130	3,893,333
2012	0	2,048,000	19,145,593	4,429,900
2013	0	8,847,400	19,458,600	11,778,271
2014	18,000	6,682,000	26,334,000	25,913,973

Data Source: The EM-DAT Database, accessed January 2015

impact of geophysical natural disasters in 2011 is significantly large and even larger than the economic impact of this type of disaster during the entire period from 1985 to 1994 (see Table 7.6). This is mainly due to the earthquake and tsunami in Japan. It is expected that the earthquake in Japan alone resulted in an economic cost as high as US$210 billion.

Selected "Green Growth" Initiatives in Asia: Renewable Energy

Asian countries have introduced several "green growth" initiatives covering many aspects such as energy efficiency, deforestation, green construction, renewable energy, and electric vehicles. This chapter focuses on selected "green growth" initiatives in renewable energy. Brunei Darussalam has no specific regulation on renewable energy development. It is a part of the general energy policy which is known as Brunei Darussalam Energy White Policy. Brunei Darussalam has a moderate target in terms of the development of renewable energy, particularly solar energy. They have set a target of increasing the share of renewable energy by 2.7 % (or about 124,000 MW) by 2017 and 10 % (954,000 MW) by 2035 (Energy Department Prime Minister's Office Brunei Darussalam 2014). Currently, Brunei Darussalam produces approximately 1700 MW of solar energy annually.

In order to promote renewable energy, Cambodia introduced the Renewable Energy Action Plan (REAP) and the Rural Electricity Master Plan (REMP). Cambodia focuses mainly on the development of hydropower electricity plants. Unfortunately there is no specific target on those two policies regarding the share of renewable energy in Cambodia's energy mix or the time frame (Poch 2013). Based on the IRENA country profile, the Government of Cambodia has announced an intention to achieve the target of 15 % of rural electricity supply from renewables by 2015 (IRENA 2013b).

Compared to other ASEAN member states, Indonesia has rich reserves of renewable energy sources particularly geothermal. Indonesia has also introduced several regulations to develop renewable energy, such as:

- Geothermal Law (Law No. 27/2003), which was introduced in 2003
- Green Energy Policy, introduced in 2004
- National Energy Blueprint, introduced in 2005
- National Biofuel Roadmap 2006–2025, introduced in 2005
- Tax rebates for investment in renewable energy (Law No. 25/2007), introduced in 2007
- Ministerial Regulation No. 32/2008 on Biofuels, introduced in 2008
- Tax incentives for geothermal energy, introduced in 2008
- Biofuels subsidy, introduced in 2009
- Tax incentive for geothermal exploration, introduced in 2011
- Feed-in tariff for geothermal electricity, introduced in 2011
- The revision of Law No. 27/2003 on geothermal which was introduced in 2014

By using various incentives, the Government of Indonesia has set several targets, namely, to (1) increase the share of renewable energy in the Indonesian energy mix up to 17 % by 2025, (2) achieve 400 MW of biomass-fired capacity and 1300 MW of hydropower capacity by 2015, (3) achieve 9.5 GW of renewable energy by 2025, and (4) achieve a 5 % blend of biodiesel and bioethanol by 2025.

Malaysia has also introduced several policies on the development of renewable energy. In 2001, the policy on fuel diversification and small REAP was implemented. Malaysia then started the development of biofuel as an alternative source of renewable energy by introducing the National Biofuel Policy in 2006. In 2009, Malaysia introduced the National Green Technology Policy which was later supported by the regulation on the financing scheme for green technology as announced in 2010. Malaysia has also introduced three comprehensive policies on renewable energy, namely, the Renewable Energy Policy and Action Plan in

2010, the Renewable Energy Act, and the Renewable Energy Feed-in Tariff in 2011. Malaysia aims to gradually increase the share of renewable capacity and electricity generation from renewables up to 36 % and 15 %, respectively, by 2050. Unlike Indonesia and Malaysia, Myanmar has no specific policies and no detailed targets on the development of renewable energy. The policies on renewable energy form part of the national economic policy, hence the National Sustainable Development Strategies announced in 2009.

The Philippines started their renewable energy program in 1977. The Philippines is the pioneer of geothermal energy in Southeast Asia and became the second biggest geothermal producer in 2000 after the USA. The comprehensive policy on renewable energy, the New and Renewable Energy Sources Development Program, was firstly announced in 1997. Then, the Philippines released its national REAP 2010–2030 in 2011. The Philippines focuses on several types of renewable energy such as geothermal, biofuels, wind, solar, hydropower, biomass, and ocean. They have set optimistic targets for 2030. The Philippines aims to triple its renewable capacity by 2030 relative to 2010. Moreover, the Philippines also wants to increase the share of renewable energy in electricity generation up to 40 % by 2020.

The contribution of renewable energy in the Singaporean energy mix is relatively small. Based on Appendix A, the total primary energy supply from renewables was only 1.1 ktoe in 2012. As explained in the previous subchapter, waste is excluded in all tables in Appendix A. If waste is also considered as a type of renewable energy, the total primary energy supply from renewables in Singapore would be equal to 603.3 ktoe in 2012. As regards policies, Singapore focuses mainly on innovation and research and development (R&D) activities. In 2001, Singapore established the Innovation for Environmental Sustainability Fund. The Clean Energy Research and Testbedding Programme was then launched in 2007. A year later, Singapore created the Solar Energy Research Institute of Singapore. Next, Singapore extended their research focus onto bioenergy by introducing the Bioenergy Research Program in 2009. In 2011, Singapore established an Experimental Power Grid Center.

Thailand launched its National Renewable Energy Development Plan 2008–2022 in 2009. Thailand has been focusing on developing several types of renewables in order to achieve their target by 2022. Wind capacity and solar capacity are expected to reach 1200 MW and 2000 MW by 2022, respectively. Within the same time frame, Thailand is expected to increase the small hydropower and biomass-fired capacity up to 1608 and 3630 MW by 2022. Vietnam issued its first Renewable Energy Action Plan in 1999. Then it was revised in 2010, resulting in a new comprehensive policy on renewable energy which is known as the Renewable Energy Development Plan. Vietnam chose its northern region as the main location for the development of renewable energy. Vietnam focuses on the development of biomass, solar, and nuclear energy. The utilization of nuclear energy as one of the sources of renewable energy in Vietnam has been initiated since 1994. Vietnam established the Vietnam Agency for Radiation and Nuclear Safety in 1994. The main task of this agency is to help the Ministry of Science and Technology in the management of radiation and nuclear safety. The preparations for nuclear power plants in Vietnam are one step ahead than those of other ASEAN member states. It is expected that the construction of the first nuclear power plant will start in 2017.

Policies on the development of renewable energy in Japan and Korea are basically similar to Singapore. The focus of the Governments of Japan and Korea is mainly on R&D activities. In Japan, the Government "forces" all government office buildings to use solar power as one of their energy sources. This program has been running since 2001. Moreover, the Government of Japan also focuses on the development of biomass and wind energy. The Government of Japan and the private sector have initiated as many as 26 projects relating to the construction of wind power plants which will produce approximately 700 MW of electricity. The Government of the Republic of Korea has also launched regulations relating to the mandatory use of renewable energy in public buildings since 2011. The Government has also announced a big project on the development of tide energy which will produce as much as 820 MW by the end of 2015.

China also has big concerns regarding the development of renewable energy. The Chinese Government has introduced several economic incentives for all renewable energy producers, such as the reduction of value-added tax (VAT) for renewable energy, import duty removal on wind and hybrid equipment, and a special fund for the industrialization of wind power equipment. The projects relating to renewable energy in China cover many types of renewable energy, including biodiesel, bioethanol, biomass, solar power, wind, marine, and small hydropower.

The Government of India introduced several incentives for the development of renewable energy. In 2007 India announced a specific regulation on incentives for ethanol production. Moreover, the Government of India also provided many forms of assistance for producers of renewable energy, covering wind power, small hydropower, and biogas. Unlike India, Pakistan focuses primarily on the development of solar photovoltaic, solar thermal, and biomass.

The initiatives on renewable energy development are not only coming from individual Asian countries but also from multilateral organizations such as the ADB. ADB launched its Clean Energy Program in 2000. This program is designed as a multipronged program that will assist Asian countries in increasing their energy efficiency, adopting renewable energy, and achieving energy security. In 2008, about 27 % of the total approved loans were accompanied by clean energy components. The Clean Energy Program covers several types of renewable energy, namely, biomass, solar, wind, and hydropower.

References

ADB (2008) Strategy 2020: the long-term strategic framework of the Asian Development Bank 2008–2020. ADB, Manila

Anderson J (2006) Climate change and natural disasters: scientific evidence of a possible relation between recent natural disasters and climate change. Briefing Note, Policy Department Economy and Science DG Internal Policies, European Parliament, Brussels

Birsic RJ (1980) The Philippines geothermal success story. Geotherm Energy 8:35–44

Daiss T (2012) Going nuclear in Southeast Asia. http://www.energytribune.com/64902/going-nuclear-in-southeast-asia#sthash.10pf7EXG.2ZfDloCr.dpbs. Accessed 28 Dec 2015
Dellink R, Lanzi E, Chateu J, Bosello F, Parrado R, De Bruin K (2014) Consequences of climate change damages for economic growth: a dynamic quantitative assessment. OECD Economics Department Working Papers, No. 1135, OECD Publishing. doi:10.1787/5jz2bxb8kmf3-en
Dickson MH, Fanelli M (2004) What is geothermal energy? http://www.geothermal-energy.org/what_is_geothermal_energy.html. Accessed 28 Dec 2015
Energy Department Prime Minister's Office Brunei Darussalam (2014) Energy white paper. Energy Department Prime Minister's Office Brunei Darussalam, Bandar Seri Begawan
Finenko A (2014) The future of Singapore's renewable energy. http://asian-power.com/regulation/commentary/future-singapores-renewable-energy. Accessed 4 Jan 2015
Huppert HE, Sparks RSJ (2006) Extreme natural hazards: population growth, population growth, globalization and environmental change. Phil Trans R Soci A 373:1875–1888
IMF (2014) Regional economic outlook. Asia and Pacific. Sustaining the momentum: vigilance and reforms. IMF, Washington, DC
IRENA (2013a) Renewable energy country profile: Brunei Darussalam. http://www.irena.org/REmaps/countryprofiles/asia/Brunei.pdf#zoom=75. Accessed 4 Jan 2015
IRENA (2013b) Renewable energy country profile: Cambodia. http://www.irena.org/REmaps/countryprofiles/asia/Cambodia.pdf#zoom=75. Accessed 4 Jan 2015
Karl TR, Nichols N, Gregory J (1997) The coming climate. Sci Am 276:78–83
Mitchell JFB, Lowe JA, Wood RA, Vellinga M (2006) Extreme events due to human induced climate change. Phil Trans R Soc A 364:2117–2113
Myint TZ, Aung N (2013) Myanmar country report. In Kimura S (ed) Analysis on energy saving potential in East Asia. ERIA Research Project Report 2012-19, Jakarta, pp 225–242
Nugroho H (2011) ASEAN energy cooperation: facts and challenges. Available via the Jakarta Post. http://www.thejakartapost.com/news/2011/05/19/asean-energy-cooperation-facts-and-challenges.html. Accessed 28 Dec 2015
OECD (2014) Towards green growth in Southeast Asia. OECD green growth studies. OECD Publishing. doi:10.1787/9789264224100-en
Oxford Business Group (2013) Brunei Darussalam sharpens focus on renewable energy. http://www.oxfordbusinessgroup.com/news/brunei-darussalam-sharpens-focus-renewable-energy. Accessed 4 Jan 2015
Pitt MM (1985) Equity, externalities, and energy subsidies: the case of kerosene in Indonesia. J Dev Econ 17:201–217
Poch K (2013) Renewable energy development in Cambodia: status, prospects and policies. In Kimura S, Phoumin H, Jacobs B (eds) Energy market integration in East Asia: renewable energy and its deployment into the power system. ERIA Research Project Report 2012-26, Jakarta, pp 227–266
UN (2012) World population prospects: the 2012 revision. UN, New York
World Bank (2013) Turn down the heat: climate extreme, regional impacts, and the case of resilience. World Bank, Washington, DC
World Nuclear Association (2014) Nuclear power in Vietnam. http://www.world-nuclear.org/info/Country-Profiles/Countries-T-Z/Vietnam/. Accessed 28 Dec 2015

Policy Conclusions

8

Policymakers have an impact on sustainability through the framework conditions that are set for the private sector, through public procurement, through government policy intervention in various fields, and through impulses for double sustainability: namely, setting incentives or new standards for the long-term financing of projects which generate positive externalities for the environment or help to reduce negative environmental externalities on the one hand; on the other hand there should be an impulse for more long-term stable banking and financial services, so that there is a contribution to financial and macroeconomic stability. Investing in countries, projects, and products which represent a high level of sustainability could be useful in this context.

Green financial products have become increasingly popular in the wake of the US subprime crisis and the Transatlantic Banking Crisis, respectively. In Germany, the largest public bank, KfW, has successfully placed several "green bonds"—in the context of renewable energy projects—which carry rather low yields, and at the same time there are extra costs for the bank for certification. The bank itself is investing in green bonds in the wholesale market in 2014/2015. If leading public and private banks should adopt more green financial products, this could help contributing to a more sustainable society. At the same time, one should not be naïve: One should not easily expect high yields on green financial products; however, the volatility of such projects might be lower if high liquidity is assured and due emphasis is placed on long-term projects—carefully evaluated before the start of the project. Two key elements for the successful expansion of such green financial product innovations are missing:

- There is a general need to introduce international standards for high-quality financial innovations. A patent for such products might even be considered under certain circumstances. The quality of financial innovations so far is quite opaque, and the high risk of financial innovations—from the clients' side—means that prices are lower than in a market with a high reputation of financial innovators.

P.J.J. Welfens et al., *Towards Global Sustainability*,
DOI 10.1007/978-3-319-18666-5_8

- There is a lack of international placement of bonds in the field of renewable energy that naturally concerns the global public good "climate stability." Why would the USA, ASEAN countries, and the EU—or some EU countries—not issue joint project bonds on innovative projects for raising energy and resource efficiency or for reducing the CO_2 intensity of production and consumption?

The EIIW-vita sustainability indicator can contribute to more transparency with regard to sustainability dynamics and sustainability policy across countries (see maps in Appendices A and B; Appendix C explains the OECD list of environmental products; Appendix D gives the list of data sources; Appendix E indicates CO_2 emissions; and Appendix F shows the Hybrid-EIIW-vita index); a simple indicator-based approach, as suggested here, has several advantages: It is easy to communicate, and it should give investors interested in sustainable investment opportunities a first view across alternative international target countries. The global sustainability indicator (GSI) developed here has the advantage that it gives four incentives:

- To raise the share of renewable energy; Germany is a particular interesting country as it is moving relatively fast from fossil fuels to a very high share of renewable energy: solar energy and wind power plus water energy generation are key elements in the renewable energy strategy of Germany—add to this biomass, which can have the function of a battery since biomass power can be stored, and one has an interesting mix of renewable energy, possibly reinforced by geothermal power
- To emphasize true savings and to thereby encourage a broader perspective on long run capital formation, including human capital formation (the World Bank concept has, however, a weakness in the area of covering the role declining stocks of natural resources—only nonrenewable resources should be included here)
- To encourage a stronger specialization in the production of environmentally friendly products and exports, respectively: The OECD has developed the concept of environmentally friendly products, and it would be useful to have regular OECD reports on the state of green innovation and export dynamics in OECD countries and the BRIICS countries
- To consider more carefully the role of water productivity—clean water is a scarce resource in many regions and countries of the world economy, and increasing scarcity in some regions could mean an increasing number of conflicts in those regions

An additional key aspect concerns the fact that the indicator is not only useful for monitoring individual countries but also for assessing the global state of sustainability (see Fig. 8.1). It is clear that there is also a broader role for more research in the field of trade, foreign direct investment, and environmental policy dynamics (Bretschger and Valente 2011; Bleischwitz 2010; Bleischwitz et al. 2012).

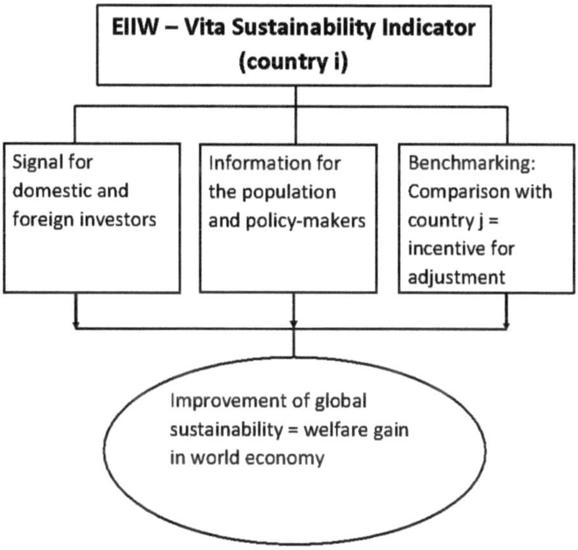

Fig. 8.1 Key policy aspects of the GSI

There is a broad international challenge for European countries and the global community, respectively. The energy sector has two particular traits which make it important from both an economic and a political perspective:

- Investment in the energy-producing sector is characterized by high capital intensity and long amortization periods, so adequate long-term planning in the private and the public sectors is required. Such long-term planning—including financing—is not available in the whole world economy, while the Transatlantic Banking Crisis has clearly undermined the stability of the international financial system and created serious problems for long-term financing. Thus, the banking crisis is directly undermining the prospects of sustainability policies in many countries.
- Investments of energy users are also mostly long term. Therefore, it takes time to switch to new, more energy-efficient consumption patterns. As energy generation and traffic accounts for almost half of global SO_2 emissions, it would be wise to not only focus on innovation in the energy sector and in energy-intensive products but also to reconsider the topic of the spatial organization of production. As long as transportation is not fully integrated into CO_2 emission certificate trading, the price of transportation is too low—negative external global warming effects are not included in market prices. This also implies that international trading patterns are often overextended. Import taxes on the weight of imported products might be a remedy to be considered by policymakers, as emissions in the transportation of goods are proportionate to the weight of the goods (actually to ton-kilometers).

One key problem for the general public as well as for policymakers is the inability of simple indicators to convey a clear message about the status of the quality of environmental and economic dynamics. The traditional system of national accounts does not provide a comprehensive approach which includes crucial green aspects of sustainability. The UN has considered several green satellite systems, but in reality the standard system of national accounts has effectively remained in place so that new impulses for global sustainability could almost be derived from standard macroeconomic figures. The GSIs presented here are a fresh approach to move towards a better understanding of the international position of countries and, hence, for the appropriate policy options to be considered in the field of sustainability policies. International organizations, governments, the general public, as well as firms could be interested in a rather simple set of indicators, which convey consistent signals for achieving a higher degree of global sustainability. The proposed indicators are a modest contribution to the international debate, and they could certainly be refined in several ways. For instance, more dimensions of green economic development might be considered, and the future path of economic and ecological dynamics might be assessed by including revealed comparative advantages (or relative world patent shares) in the field of "green patenting." The new proposed indicators could be important elements of an environmental and economic compass which would suggest optimum routes for intelligent green development. As the HDI index of the UN is negatively correlated with the EIIW-vita index, one can construct a new hybrid indicator which gives insights into green economic welfare, broadly defined.

The GSI provides broad information to firms and consumers in their respective countries and thus could encourage green innovations and new environmentally friendly consumption patterns.

The GSI also encourages governments in countries eager to catch up with leading countries to provide adequate innovation incentives for firms and households, respectively. This could in turn encourage international diffusion of best practice, thereby contributing to enhanced global sustainability in the world economy.

The Copenhagen process will show to what extent policymakers and actors in the business community are able to find new international solutions and to set the right incentives for more innovations in the climate policy arena. There is no reason to be pessimistic; on the contrary, with a worldwide common interest in controlling global warming, there is a new field which might trigger more useful international cooperation among policymakers, in general, and among environmental policies, in particular. From an innovation policy perspective, there is, however, some cause for pessimism in the sense that the Old Economy industries—most of them are highly energy intensive—are well established and have strong links to the political system, while small- and medium-sized innovative firms with the relevant R&D activities in global climate control typically find it very difficult to get political support. Thus, one should consider imposing specific taxes on nonrenewable energy producers and use the proceeds to largely stimulate green innovative firms and sectors, respectively. Competition, free trade, and foreign direct investment all have their role to play in technology diffusion, but without a critical minimum effort by the EU, Switzerland, Norway, the USA, China, India, the ASEAN countries, and many others, it is not realistic to assume that a radical reduction of CO_2 emissions can be achieved by 2050. Emphasis should also be put on restoring stability in the financial sector and encouraging banks and other financial institutions to take a more long-term view. Here it would be useful to adopt a volatility tax which would be imposed on the variance (or the coefficient of variation) of the rate of return on equity of banks (Welfens 2009).

It is yet to be seen whether or not the Rio Process can deliver meaningful results in the medium-term and in the long run. If the financial sector in OECD countries and elsewhere remains in a shaky condition, long-term financing for investment and innovation will be difficult to obtain in the marketplace. This brings us back to the initial conjecture that we need double sustainability—i.e., both in the banking sector and in the overall economy. The challenges are tough, and the waters on the way to a sustainable global economic–environmental equilibrium might be rough, but the necessary instruments are known: to achieve a critical minimum of green innovation dynamics will require a careful watching of standard environmental and economic statistics, and it will also be quite useful to study the results and implications of the EIIW-vita GSI.

The EIIW-vita Sustainability Indicator can generate adequate modernization impulses for all economies of the world in which people want to achieve sustainable development. The results from the indicator show both the respective status and evolution over time; however, it does not automatically indicate which adjustment measures should be taken. Here theoretical reflections, empirical findings, and the institutional constraints should be taken into account. In some cases, optimum growth theory—the golden rule—could be helpful as a complementary analytical framework, particularly if the capital intensity is too high compared to the ratio of capital stock to worker: The golden rule allows to maximize long run per capita consumption, and a country whose capital intensity (K/L) exceeds the golden rule capital intensity is not only failing to achieve maximum sustainable per capita consumption but also has an excessive capital stock whose production and use will generate more emissions (in a life-cycle perspective, the energy used during the production of the respective machinery and equipment also has to be included) than necessary (Welfens 2014). In some countries, such a problem could indeed be relevant. From this perspective, economic optimization is, of course, an element of minimizing the use of scarce resources. Efficiency is almost always a contribution to sustainability; however, this is a necessary condition for sustainability but not a sufficient condition to achieve sustainable development.

Organizational Implications for Sustainability

Global economic growth and increasing climate problems are occurring simultaneously. Investors, households, and policymakers all have an impact on the degree to which growth will become greener and thus more sustainable. Individuals face a specific responsibility when buying products, i.e., to consider recycling options and giving a second lease of life to products by purchasing used durable goods over the internet. There is also an individual influence and responsibility when one considers investment decisions, as investments in certain sectors and products stimulate a specific structural change. Investors often have no adequate information about the sustainability indicators of individual companies or of the effective degree of sustainability represented by investment funds—when it comes to a country-related investment fund, the EIIW-vita Sustainability Indicator gives explicit information which could and should be considered by private investors. Countries in the upper half of the EVSI league as well as those countries achieving fast progress in the field of sustainability should be carefully considered as interesting investment opportunities.

As regards policymakers, one should consider at least four crucial layers:

- The UN has a crucial role in organizing and coordinating the fight of countries against global warming.
- The G20 is a rather new grouping of countries that has gained in importance since late 2008 when the US banking crisis effectively forced the old G8 group to organize stabilization and rescue measures within a broader forum, namely, the G20. The G20 is much more compact than the UN, with its almost 200-member countries; however, the G20 is also very heterogeneous, as the per capita income in 2013 was in the range of US$54,678 in the United States to 5777 in India (expressed in purchasing power parity). The large income discrepancies imply that interests across countries are rather heterogeneous. In economic terms, the G20 is a group of countries dominated by Asia, and hence the USA and the EU might have some reservations about this forum.
- There are supranational institutions such as the European Union or Mercosur—both customs unions (with common external tariffs)—or NAFTA, ASEAN, or the Gulf Cooperation Council, etc., which stand for joint regional policy. To the extent that big actors compete in the world economy or are implementing cooperative policy approaches, regional integration can be key to more global sustainability as such integration can use the political interaction costs in the world economy: If simply several big groups are negotiating, for example, at climate policy meetings, it should be easier to achieve a compromise and a consensus, respectively—compared to a situation in which 200 countries negotiate about environmental issues at the UN. Foreign investors and policy partners are increasingly considering the fact that it is not only individual countries which are active in various policy fields but that regional integration schemes represent some specific form of cooperation, occasionally for a more ambitious or more efficient form of environmental policy. Industrial standards as well as policy standards may be implemented in certain regional integration schemes and more sustainability may be achieved jointly than in the case of individual, uncoordinated actions. We show subsequently the maps for the EU and ASEAN in 2010, and one can indeed see that there are considerable differences—often in favor of Europe.

- National policymakers play a decisive role. The more the general public is interested in sustainability indicators, the more one will demand to learn about the relative positioning of one's own country in comparison with other countries. As regards the USA and the BRIICS, each of these seven "super-big" countries has a decisive impact on the world climate by itself. Moreover, it should not be too difficult to organize efficient coordination among a small group of big countries.

The policy conclusion clearly is that each policy layer should assume its responsibility. There is a problem in the sense that the UN environmental approach, and that of the G20, is rather bureaucratic. The G20 has produced many useful ideas and suggestions; however, the amount of truly implemented policy reforms, for example, in the field of cutting energy subsidies—effectively a fossil fuel subsidy in many developing countries and an atomic energy subsidy in OECD countries—is rather modest. The proposals of the Washington G20 Communiqué have not turned out to be, politically speaking, very influential. The G20 as a "policy club of countries" is a heterogeneous group for organizing political consensus as differences in per capita income are enormous, and hence there are strong conflicts of interest. At the same time, it holds that the old G8 form of leadership is no longer working and has lost its once powerful status—it was in the wake of the US banking crisis that the G20 was activated at the end of 2008, and it will take some time for the G20 activities to be better organized. One may doubt that the current structure, with the IMF being used as a quasi-secretariat for the G20, is an efficient institutional architecture.

It would be quite interesting if regional integration clubs could cooperate. For example, if the EU, ASEAN, Mercosur, NAFTA, the Gulf Cooperation Council, Russia, India, Indonesia, and China would cooperate in a G9 framework, one would have broader coverage than G20—in terms of the share of world GDP, the share of global emissions and the share of global population; at the same time, it would be a much more compact and easier to organize group.

The bottom line is that we should like to encourage a broad international debate about composite sustainability indicators and the policy reforms needed to achieve sustainability for both North and South. We consider a broader indicator which includes water productivity—and possibly also recycling—as a useful extension of earlier work and will conduct further research.

References

Bleischwitz R (2010) International economics of resource productivity: relevance, measurement, empirical trends, innovation, resource policies. IEEP 7(2–3):227–244

Bleischwitz R, Welfens PJJ, Zhang Z (2012) Prologue to the special issue "Resources, energy and eco-innovation". Miner Econ 24:55–57

Bretschger L, Valente S (2011) International trade and net investment: theory and evidence. IEEP 8(2):197–224

Welfens PJJ (2009) Transatlantische Bankenkrise. Lucius & Lucius, Stuttgart
Welfens PJJ (2014) Green innovations and CO_2 emissions in a growth perspective: competition, growth, welfare analysis and policy implications. In: Meckl R, Rongping M (eds) Innovation for green growth. Science Press, Beijing, pp 19–40

Appendix

Appendix A: World Maps, 2000–2011 (Dark Green Is Best)

2000

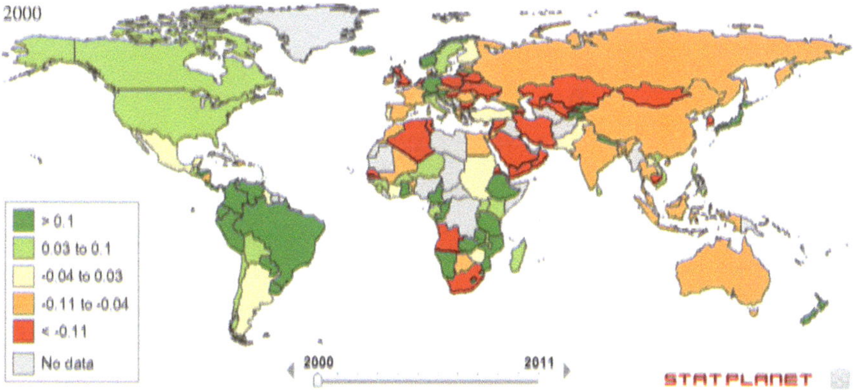

2001

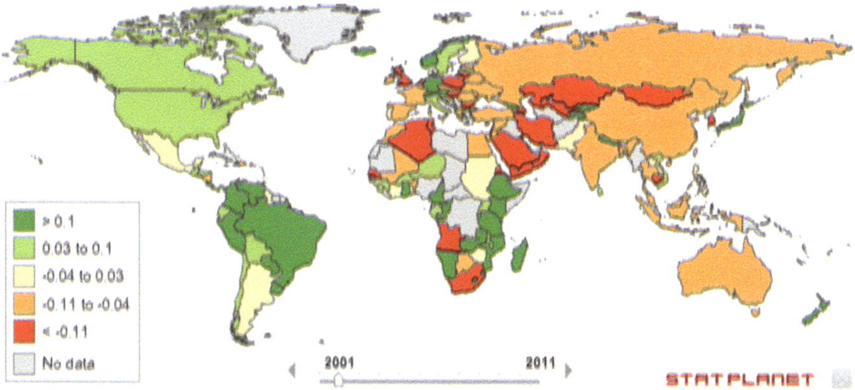

P.J.J. Welfens et al., *Towards Global Sustainability*,
DOI 10.1007/978-3-319-18666-5

2002

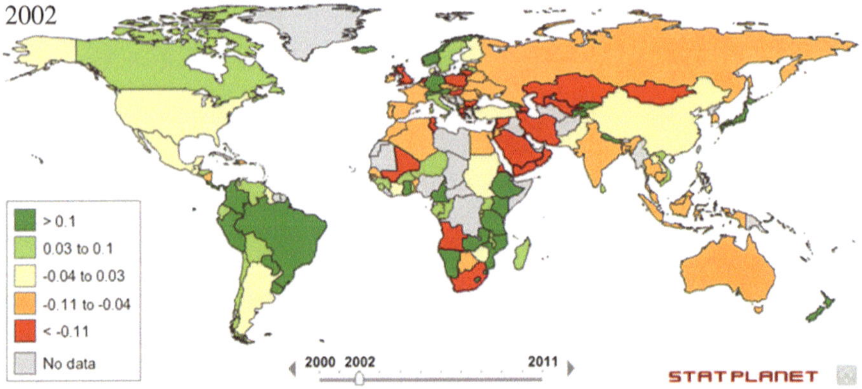

2003

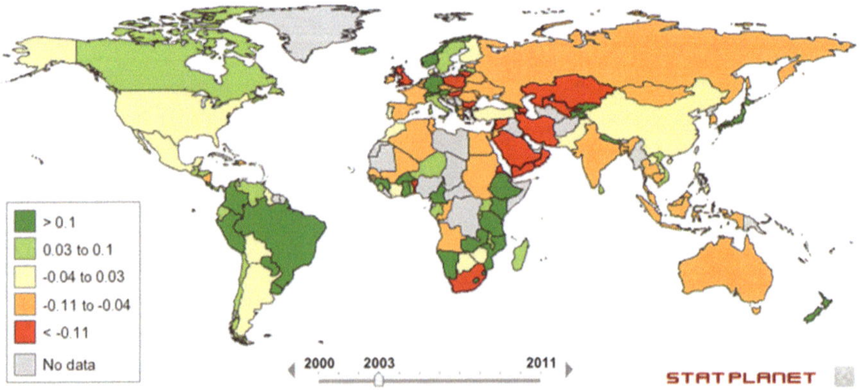

2004

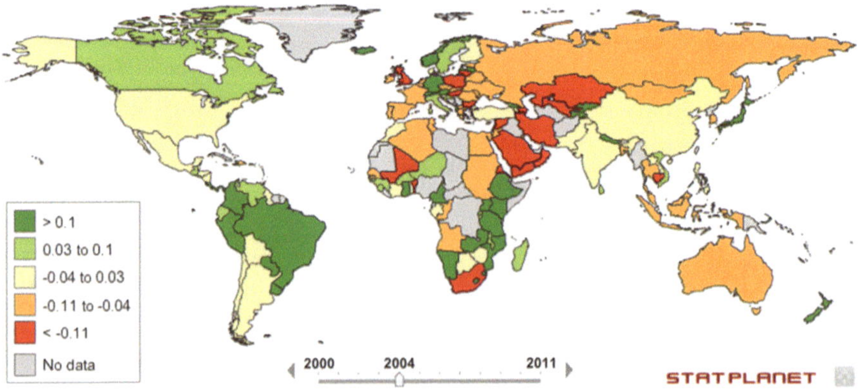

2005

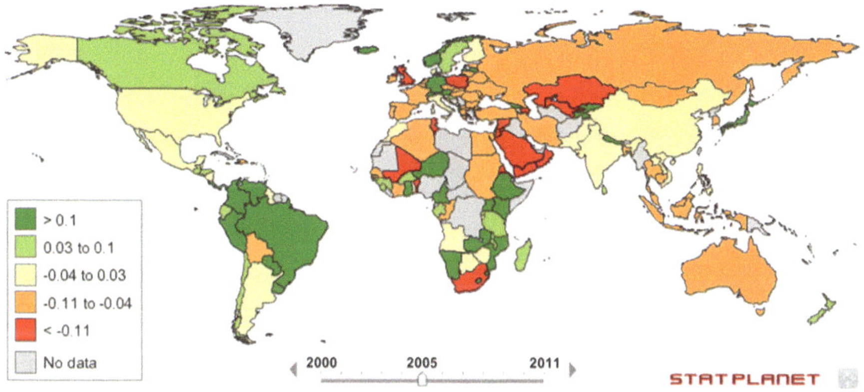

2006

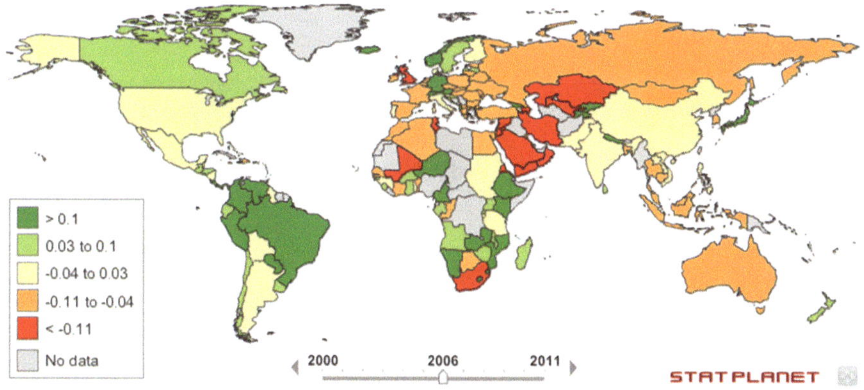

2007

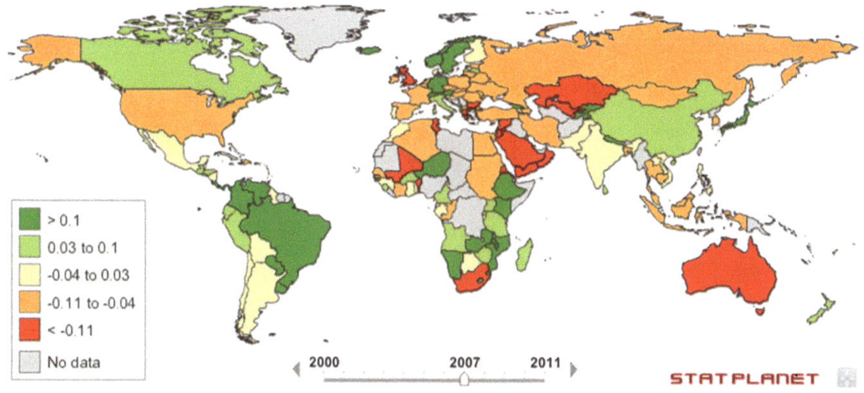

2008

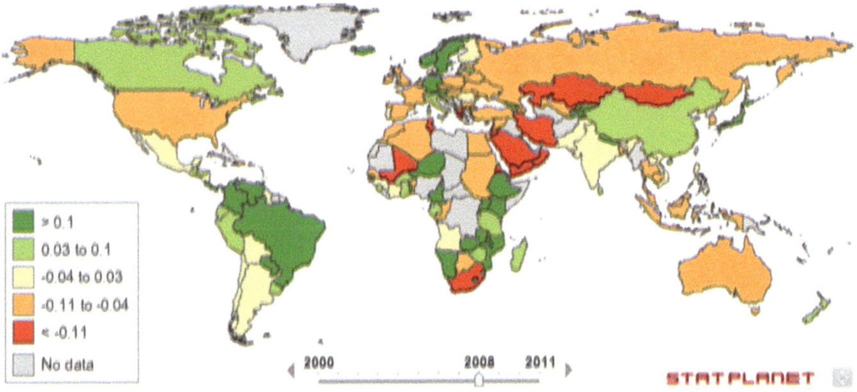

2009

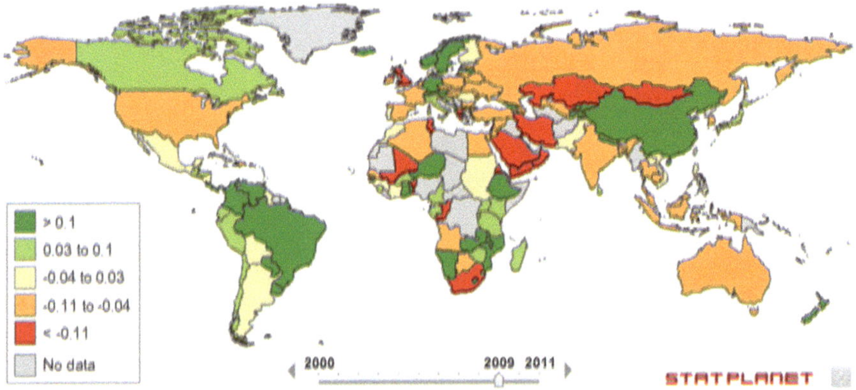

2010

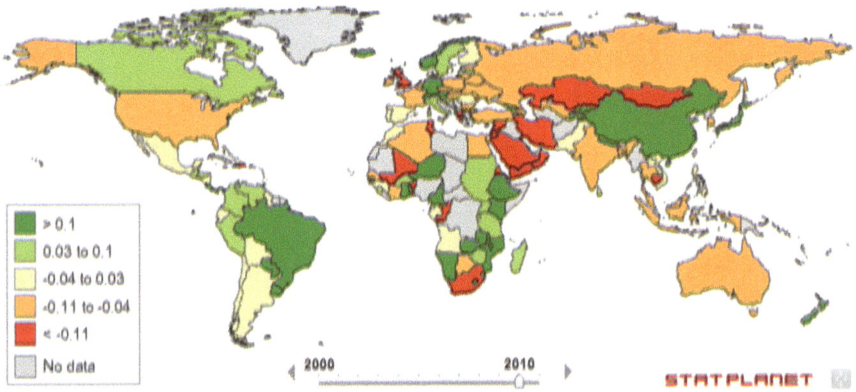

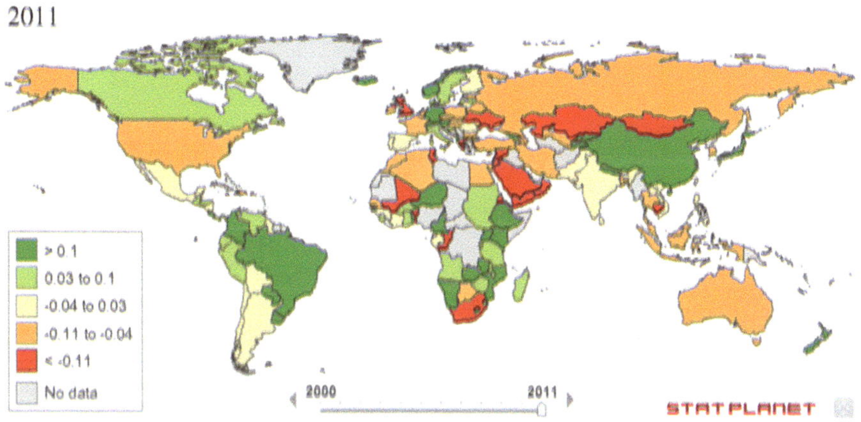

Source: EIIW calculations based on data from the World Bank, World Development Indicators and UN Comtrade, DESA/UNSD. StatSilk (2015) StatPlanet: Interactive Data Visualization and Mapping Software. http://www.statsilk.com

Appendix B: EIIW-vita Indicator for Europe, 2000 and 2011

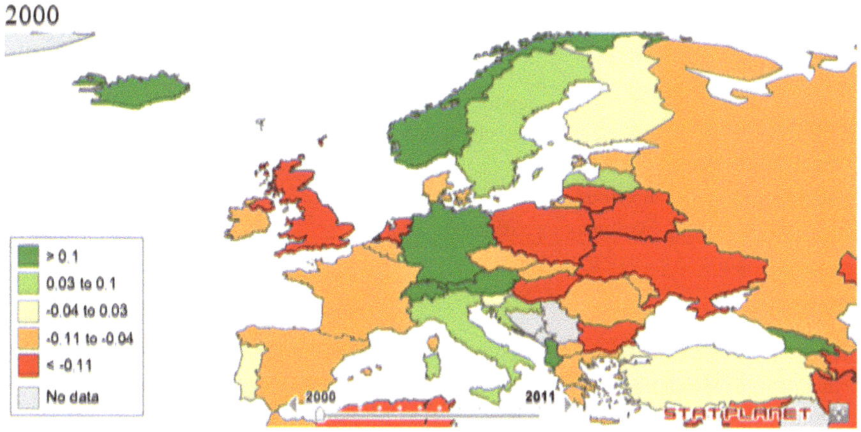

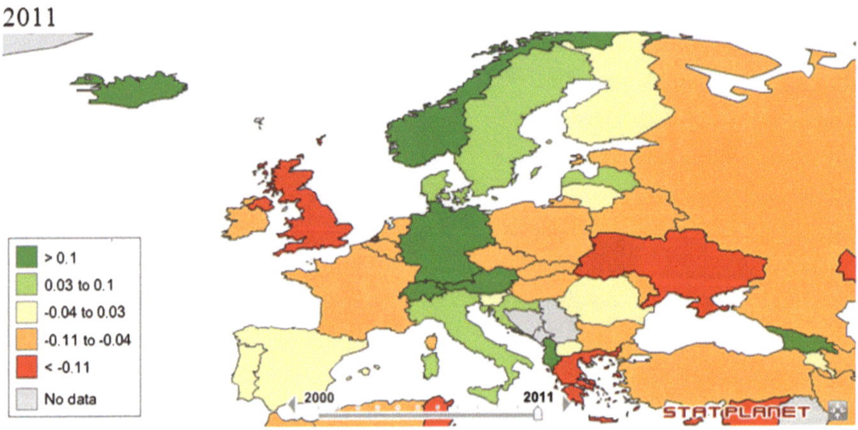

Source: EIIW calculations based on data from the World Bank, World Development Indicators and UN Comtrade, DESA/UNSD. StatSilk (2015) StatPlanet: Interactive Data Visualization and Mapping Software. http://www.statsilk.com

Appendix C: List of Environmental Products (OECD)

Description	HS
Vacuum pumps	841410
Compressors of a kind used in refrigerating equipment	841430
Air compressors mounted on a wheeled chassis for towing	841440
Other air or gas compressors or hoods	841480
Parts for air or gas compressors, fans, or hoods	841490
Filtering or purifying machinery and apparatus for gases	842139
Parts for filtering or purifying machinery	842199
Limestone flux	252100
Slaked (hydrated) lime	252220
Magnesium hydroxide and peroxide activated earths	281610
Filtering or purifying machinery and apparatus for gases	842139
Parts for filtering or purifying machinery	842199
Filtering or purifying machinery and apparatus for gases	842139
Parts for filtering or purifying machinery	842199
Other glass fiber products	701990
Machinery for liquefying air or other gases	841960
Other machinery, for treatment of materials by change of Temperature	841989
Filtering or purifying machinery and apparatus for gases	842139
Parts for filtering or purifying machinery	842199
Other furnaces, ovens, incinerators, nonelectric	841780
Filtering or purifying machinery and apparatus for gases	842139

(continued)

Description	HS
Parts for filtering or purifying machinery	842199
Industrial or laboratory electric resistance furnaces	851410
Industrial or laboratory induction or dielectric furnaces	851420
Other industrial or laboratory electric furnaces and ovens	851430
Parts, industrial or laboratory electric furnaces	851490
Parts for sprayers for powders or liquids	842490
Compressors of a kind used in refrigerating equipment	841430
Air compressors mounted on a wheeled chassis for towing	841440
Other air or gas compressors or hoods	841480
Parts for air or gas compressors, fans, or hoods	841490
Limestone flux	252100
Phosphates of mono or disodium	283522
Phosphates of trisodium	283523
Phosphates of potassium	283524
Calcium hydrogen orthophosphate	283525
Other phosphates of calcium	283526
Other phosphates (excl. polyphosphates)	283529
Activated carbon	380210
Water filtering or purifying machinery and apparatus	842121
Other machinery for purifying liquids	842129
Parts for filtering or purifying machinery	842199
Other centrifuges	842119
Parts of centrifuges	842191
Water filtering or purifying machinery and apparatus	842121
Other machinery for purifying liquids	842129
Parts for filtering or purifying machinery	842199
Other articles of plastic	392690
Water filtering or purifying machinery and apparatus	842121
Other machinery for purifying liquids	842129
Parts for filtering or purifying machinery	842199
Woven pile and chenille fabrics of other textile materials	580190
Tanks, vats, etc. >300 l	730900
Tanks, drums, etc. >50 l < 300 l	731010
Cans <50 l, closed by soldering or crimping	731021
Other cans <50 l	731029
Hydraulic turbines	841011
	841012
	841013
Parts for hydraulic turbines	841090
Other furnaces, ovens, incinerators, nonelectric	841780
Weighing machines capacity <30 kg	842381
Weighing machines capacity >30 kg < 5000 kg	842382
Parts for sprayers for powders or liquids	842490
Industrial or laboratory electric resistance furnaces	851410

(continued)

Description	HS
Industrial or laboratory induction or dielectric furnaces	851420
Other industrial or laboratory electric furnaces and ovens	851430
Parts, industrial or laboratory electric furnaces	851430
Cast articles of cast iron	732510
Positive displacement pumps, hand operated	841320
Other reciprocating positive displacement pumps	841350
Other rotary positive displacement pumps	841360
Other centrifugal pumps	841370
Other pumps	841381
Valves, pressure reducing	848110
Valves, check	848130
Valves, safety	848140
Other taps, cocks, valves, etc.	848180
Instruments for measuring the flow or level of liquids	902610
Instruments for measuring or checking pressure	902620
Other articles of cement, concrete	681099
Other articles of lead	780600
Other electric space heating and soil heating apparatus	851629
Lasers Vitrification equipment	901320
Household or toilet articles of plastic	392490
Brooms, hand	960310
Brushes as parts of machines, appliances	960350
Mechanical floor sweepers Trash bin liners (plastic)	960390
Polypropylene sheeting, etc.	392020
Machinery to clean, dry bottles, etc.	842220
Other mixing or kneading machines for earth, stone, sand, etc.	847439
Other machines for mixing/grinding, etc.	847982
Other machines, having individual functions	847989
Other furnaces, ovens, incinerators, nonelectric	841780
Parts of furnaces, nonelectric	841790
Industrial or laboratory electric resistance furnaces	851410
Industrial or laboratory induction or dielectric furnaces	851420
Other industrial or laboratory electric furnaces and ovens	851430
Parts, industrial or laboratory electric furnaces	851490
Cleaning-up	851629
Other electric space heating and soil heating apparatus	901320
Other electrical machines and apparatus with one function	854389
Parts for spark-ignition internal combustion piston engines	840991
Parts for diesel or semi-diesel engines	840999
Silencers and exhaust pipes, motor vehicles	870892
Thermometers, pyrometers, liquid filled	902511
Other thermometers, pyrometers	902519
Hydrometers, barometers, hygrometers, etc.	902580
Other instruments for measuring liquids or gases	902680
Parts of instruments for measuring, checking liquids or gases	902690

(continued)

Description	HS
Instruments for analysing gas or smoke	902710
Chromatographs, etc.	902720
Spectrometers, etc.	902730
Exposure meters	902740
Other instruments using optical radiation	902750
Other instruments for physical or chemical analysis	902780
Parts for instruments, incl. microtomes	902790
Ionising radiation measuring or detecting instruments	903010
Other optical instruments	903149
Other measuring or checking instruments	903180
Manostats	903220
Hydraulic/pneumatic automatic regulating, controlling instruments	903281
Other automatic regulating, controlling instruments auto emissions testers noise measuring equipment	903289
Thermostats	903210
Peat replacements (e.g., bark)	284700
Paints and varnishes, in aqueous medium, acrylic, or vinyl	320910
Other paints and varnishes, in aqueous medium	320990
Chlorine	280110
Waters, including natural or artificial mineral water	220100
Distilled and conductivity water	285100
Ion exchangers (polymer)	391400
Instantaneous gas water heaters	841911
Other instantaneous or storage water heaters, nonelectric	841919
Photosensitive semiconductor devices, including solar cells	854140
Methanol	290511
Multiple walled insulating units of glass	700800
Other glass fibre products	701990
Heat exchange units	841950
Parts for heat exchange equipment	841990
Fluorescent lamps, hot cathode	853931
Gas supply, production and calibrating meters	902810
Liquid supply, production and calibrating meters	902820
Thermostats	903210

Appendix D: List of Data Sources

Source	Data
UN Comtrade, DESA/UNSD	International Trade Data
World Bank, World Development Indicators Database	National Data for Savings, Share of Renewables, Total Exports, CO_2 Emissions

(continued)

Source	Data
OECD Manual for Environmental Goods	List of Environmental Products

Appendix E: CO_2 Emissions (kg CO_2 per 1000 US Dollar)

	1990	2005	2013
Afghanistan	85.49	74.53	13.35
Albania	319.95	364.10	70.42
Algeria	151.40	175.61	157.29
Angola	224.40	145.60	93.33
Antigua and Barbuda	263.32	186.92	182.31
Argentina	382.20	275.44	212.81
Armenia	383.37	889.84	125.33
Australia	357.11	321.81	314.09
Austria	263.31	174.09	189.79
Azerbaijan	352.45	752.73	180.44
Bahamas	706.57	367.37	411.24
Bahrain	259.43	318.59	239.40
Bangladesh	33.61	47.44	51.02
Barbados	198.65	243.96	206.69
Belarus	947.90	1020.86	380.60
Belgium	484.16	274.02	288.83
Belize	192.00	71.63	87.92
Benin	71.24	36.43	61.70
Bermuda	102.09	112.27	66.57
Bhutan	41.54	41.93	57.89
Bolivia	144.66	125.59	148.19
Bosnia and Herzegovina	3482.17	747.90	130.08
Botswana	46.65	59.73	52.39
Brazil	88.70	99.30	99.31
Brunei Darussalam	120.39	120.63	175.41
Bulgaria	840.20	930.88	519.97
Burkina Faso	35.63	33.47	26.66
Burundi	17.07	57.34	53.86
Cambodia	8.31	21.06	15.10
Cameroon	45.65	101.29	100.53
Canada	461.37	350.95	334.81
Cape Verde	259.14	35.91	29.31
Central African Republic	67.30	95.70	115.93
Chad	17.61	6.86	5.06
Chile	231.08	99.49	133.87

(continued)

	1990	2005	2013
China	599.07	323.21	229.03
Colombia	114.76	122.66	108.46
Comoros	34.32	59.34	62.81
Congo	228.32	130.33	121.75
Congo, DR	68.75	129.71	46.65
Costa Rica	81.39	55.39	72.05
Cote d'Ivoire	82.36	71.27	77.51
Croatia	369.76	392.54	196.83
Cuba	99.50	141.70	74.52
Cyprus	211.27	177.27	247.90
Czech Republic	961.63	757.50	477.75
Denmark	352.53	249.71	316.45
Djibouti	489.20	369.36	441.18
Dominica	62.98	113.72	115.08
Dominican Republic	146.17	118.86	112.19
Ecuador	68.83	131.84	135.61
Egypt	83.50	135.63	117.31
El Salvador	85.42	58.39	96.31
Equatorial Guinea	92.79	3.55	50.63
Eritrea	49.07	33.42	121.93
Estonia	3708.41	1770.91	522.20
Ethiopia	33.19	28.47	24.75
Fiji	134.36	153.05	150.13
Finland	354.37	281.72	324.89
France	309.71	165.53	168.76
Gabon	152.06	170.62	186.58
The Gambia	64.45	83.87	58.46
Georgia	1210.62	1876.75	205.81
Germany	472.30	351.00	277.03
Ghana	92.14	53.51	44.80
Gibraltar	88.35	149.78	295.94
Greece	188.17	222.25	312.99
Grenada	53.68	63.33	131.87
Guatemala	61.18	42.71	61.19
Guinea	86.95	101.22	74.68
Guinea-Bissau	32.07	110.83	116.08
Guyana	586.68	300.67	191.67
Haiti	37.22	78.26	75.16
Honduras	89.20	82.89	111.82
Hong Kong	64.62	117.51	99.58
Hungary	420.61	368.97	285.06
Iceland	349.46	196.85	214.34
India	152.72	154.82	139.47
Indonesia	53.49	80.62	100.48

(continued)

	1990	2005	2013
Iran, Islamic Republic of	235.09	178.51	247.40
Iraq	162.42	232.34	157.83
Ireland	306.61	158.03	187.01
Israel	183.16	182.30	203.70
Italy	218.03	199.63	214.50
Jamaica	264.83	123.46	217.40
Japan	270.69	242.25	279.28
Jordan	90.00	184.15	194.41
Kazakhstan	533.89	575.76	413.91
Kenya	103.15	104.09	90.05
Kiribati	70.71	148.59	128.26
Korea, DPR	1289.79	2982.98	1691.36
Korea, Republic of	142.54	171.38	260.70
Kuwait	216.02	219.57	165.25
Kyrgyzstan	658.00	648.83	342.82
Lao PDR	24.52	13.66	13.82
Latvia	1070.69	1445.21	210.04
Lebanon	357.76	134.74	204.93
Lesotho	43.92	29.57	26.63
Liberia	1082.53	787.90	85.71
Libyan Arab Jamahiriya	207.27	270.10	382.81
Lithuania	1658.40	2089.94	210.37
Luxembourg	850.29	255.43	193.21
Macao	28.21	27.81	15.86
Macedonia	508.02	477.31	414.23
Madagascar	37.68	49.95	33.34
Malawi	63.41	52.33	51.15
Malaysia	80.71	88.36	149.93
Maldives	6.86	38.37	76.81
Mali	29.67	13.45	16.95
Malta	138.46	206.04	188.84
Mauritania	86.36	124.50	235.09
Mauritius	57.70	72.07	66.37
Mexico	129.91	166.90	168.50
Moldova, Republic of	1096.65	4208.62	734.50
Mongolia	152.20	1148.28	466.75
Morocco	89.54	124.05	121.69
Mozambique	534.58	101.57	42.16
Myanmar	300.98	95.00	79.06
Namibia	41.53	82.51	67.47
Nepal	21.37	21.54	35.27
Netherlands	346.20	236.28	258.22
Netherlands Antilles	5948.69	1132.43	906.89
New Zealand	235.20	171.04	185.91

(continued)

	1990	2005	2013
Nicaragua	150.86	104.98	107.32
Niger	44.99	68.13	40.78
Nigeria	167.32	110.67	97.04
Norway	213.86	131.35	135.06
Oman	124.47	115.59	123.61
Pakistan	60.37	92.31	107.63
Panama	168.28	63.63	60.72
Papua New Guinea	190.37	226.97	124.53
Paraguay	31.42	58.98	73.30
Peru	167.20	112.84	82.13
Philippines	121.27	89.78	113.25
Poland	953.67	615.57	348.08
Portugal	104.76	125.68	197.09
Puerto Rico	6.61	2.31	2.88
Qatar	304.86	163.98	81.61
Romania	541.81	727.16	371.08
Russian Federation	562.98	866.17	507.85
Rwanda	14.41	68.78	35.73
Saint Kitts and Nevis	72.57	65.04	84.00
Saint Lucia	118.45	140.20	124.86
Saint Vincent and the G.	61.97	64.66	120.00
Samoa	65.18	148.11	147.24
Sao Tome and Principe	75.97	99.16	113.71
Saudi Arabia	167.42	167.51	151.23
Senegal	111.04	92.33	96.98
Serbia and Montenegro	253.34	666.53	537.47
Seychelles	134.72	281.75	174.16
Sierra Leone	78.32	169.56	65.98
Singapore	85.38	79.83	101.11
Slovakia	565.11	599.96	326.45
Slovenia	251.41	311.08	287.06
Solomon Islands	76.69	227.86	143.71
Somalia	116.74	199.53	84.35
South Africa	616.51	509.67	464.23
Spain	176.42	143.69	168.82
Sri Lanka	69.51	38.21	45.28
Sudan	99.75	41.12	36.04
Suriname	497.49	228.84	166.89
Swaziland	64.60	74.53	62.88
Sweden	360.29	170.81	167.60
Switzerland	156.94	121.29	107.66
Syrian Arab Republic	183.49	424.71	478.65
Taiwan	162.38	186.45	229.33
Tajikistan	158.49	294.09	314.62

(continued)

	1990	2005	2013
Tanzania	74.38	40.60	42.13
Thailand	68.60	89.75	193.87
Timor-Leste	1.15	0.75	0.44
Togo	72.48	104.41	125.00
Tonga	47.15	84.74	137.93
Trinidad and Tobago	396.73	360.41	347.13
Tunisia	107.08	151.90	151.87
Turkey	107.46	121.22	138.02
Turkmenistan	645.42	455.04	454.95
Tuvalu	7.86	8.12	7.56
Uganda	41.10	25.61	15.64
Ukraine	853.93	1633.87	1011.83
United Arab Emirates	164.55	119.81	149.05
United Kingdom	499.20	278.10	254.11
United States	531.18	343.04	334.43
Uruguay	185.21	108.04	85.83
Uzbekistan	1067.42	669.04	677.31
Vanuatu	132.80	245.63	85.84
Venezuela	239.90	250.38	258.50
Viet Nam	192.84	83.24	81.74
Yemen	54.32	76.70	118.12
Zambia	227.72	128.87	43.34
Zimbabwe	278.98	555.13	539.11
EU-28	375.94	265.73	222.11
World	489.11	395.49	357.83

Note: Global CO_2 emissions from fossil fuel use and cement production per 1000 dollar GDI for each country (kg CO_2 per 1000 US dollar and per capita. US dollar is adjusted to the Purchasing Power Parity of 2011).

Source: European Commission EDGAR Database

Appendix F: Hybrid EIIW-vita-HDI Index

Country	2000	2005	2006	2007	2008	2009	2010	2011
Albania	0.4524	0.4649	0.4659	0.4679	0.4735	0.4688	0.4667	0.4658
Algeria	0.2526	0.2915	0.2944	0.3135	0.3087	0.3197	0.3126	0.3194
Angola	0.0997	0.2281	0.2719	0.2567	0.2312	0.2183	0.2484	0.2606
Argentina	0.3633	0.3731	0.3950	0.3841	0.3776	0.3863	0.3847	0.3838
Armenia	0.2825	0.3452	0.3614	0.3698	0.3689	0.3554	0.3603	0.3540
Australia	0.4020	0.4058	0.4085	0.4044	0.4075	0.4171	0.4100	0.4143
Austria	0.4887	0.4923	0.4996	0.5130	0.5161	0.5122	0.5029	0.5055
Azerbaijan							0.3231	0.3105

(continued)

Country	2000	2005	2006	2007	2008	2009	2010	2011
Bahrain	0.3163	0.3360	0.3424	0.3539	0.3515	0.3462	0.3434	0.3442
Bangladesh	0.1655	0.1939	0.1977	0.2087	0.2094	0.2138	0.2161	0.2153
Belarus		0.3256	0.3232	0.3289	0.3332	0.3261	0.3282	0.3301
Belgium	0.3863	0.3932	0.3947	0.4022	0.4027	0.3960	0.4006	0.4094
Belize	0.3510	0.3368	0.3627	0.3602	0.3693	0.3647	0.3643	0.3667
Benin	0.1360	0.1474	0.1523	0.1511	0.1569	0.1528	0.1518	0.1534
Bolivia	0.3254	0.3026	0.3187	0.3147	0.3197	0.3308	0.3221	0.3290
Botswana	0.2563	0.2924	0.2810	0.2970	0.2851	0.2724	0.2723	0.2759
Brazil	0.4189	0.4279	0.4297	0.4334	0.4311	0.4378	0.4311	0.4362
Brunei Darussalam	0.3327	0.3315	0.3442	0.3395	0.3519	0.3621	0.3511	0.3536
Bulgaria	0.3025	0.3229	0.3226	0.3043	0.3284	0.3377	0.3444	0.3451
Burkina Faso		0.1985	0.1870	0.2014	0.2094	0.2091	0.2114	0.2138
Burundi	0.1432	0.1471	0.1539	0.1393	0.1592	0.1818	0.1714	0.1864
Cambodia	0.1582	0.1924	0.2027	0.2006	0.2078	0.2041	0.2022	0.2053
Cameroon	0.3117	0.3054	0.3027	0.2788	0.2872	0.2838	0.2895	0.2943
Canada	0.4723	0.4833	0.4858	0.4858	0.4885	0.4858	0.4809	0.4855
Chile	0.3934	0.4180	0.4153	0.3984	0.4068	0.4167	0.4017	0.4011
China	0.2631	0.3154	0.3212	0.3464	0.3701	0.3913	0.4135	0.4022
Colombia	0.3856	0.4038	0.4102	0.4124	0.4207	0.4080	0.4030	0.4219
Congo, Rep.	0.3082	0.2149	0.2044	0.2201	0.2106	0.1859	0.1696	0.1884
Costa Rica	0.4597	0.4752	0.4743	0.4784	0.4783	0.4794	0.4773	0.4644
Cote d'Ivoire	0.1860	0.1622	0.1679	0.1681	0.1833	0.1973	0.1804	0.1842
Croatia	0.4033	0.4179	0.4168	0.3964	0.4178	0.4307	0.4402	0.4142
Cyprus	0.3413	0.3485	0.3498	0.3435	0.3487	0.3563	0.3552	0.3595
Czech Republic	0.3548	0.3841	0.3895	0.3921	0.3992	0.3893	0.3867	0.3941
Denmark	0.4077	0.4475	0.4331	0.4467	0.4478	0.4428	0.4501	0.4679
Dominican Republic	0.2764	0.2933	0.2931	0.2870	0.2852	0.2911	0.2870	0.2905
Ecuador	0.3895	0.3647	0.3715	0.3791	0.3961	0.3789	0.3650	0.3801
Egypt, Arab Rep.	0.2626	0.2558	0.2641	0.2713	0.2714	0.2729	0.2696	0.2671
El Salvador	0.3440	0.3568	0.3596	0.3547	0.3680	0.3629	0.3740	0.3761
Eritrea							0.0950	0.0988
Estonia	0.3332	0.3658	0.3677	0.3724	0.3691	0.3696	0.3722	0.3828
Ethiopia	0.2364	0.2510	0.2568	0.2731	0.2610	0.2627	0.2911	0.2993
Fiji	0.3058	0.3122	0.2926	0.2964	0.2979	0.2942	0.2895	0.2895
Finland	0.4222	0.4473	0.4368	0.4480	0.4561	0.4345	0.4332	0.4407
France	0.3897	0.4038	0.4077	0.4119	0.4138	0.4069	0.4033	0.4017
Gabon	0.3477	0.3419	0.3518	0.3348	0.3474	0.3478	0.3430	0.3508
The Gambia	0.2263	0.2365	0.2517	0.2436	0.2369	0.2551	0.2523	0.2557
Georgia		0.4467	0.4167	0.4300	0.4248	0.4254	0.4482	0.4305
Germany	0.4976	0.5675	0.5719	0.5863	0.5853	0.5796	0.5803	0.5904

(continued)

Country	2000	2005	2006	2007	2008	2009	2010	2011
Ghana	0.3167	0.3241	0.2913	0.2678	0.3073	0.3359	0.3235	0.3022
Greece	0.3507	0.3765	0.3842	0.3665	0.3706	0.3723	0.3747	0.3733
Guatemala	0.2845	0.3203	0.3148	0.3125	0.3176	0.3154	0.3294	0.3279
Guinea		0.1930	0.1803	0.1595	0.1766	0.1693	0.1642	0.1711
Guinea-Bissau		0.2151	0.2096	0.2516	0.2285	0.2478	0.2542	0.2616
Guyana	0.2991	0.3075	0.3202	0.3073	0.3084	0.3176	0.3144	0.3194
Haiti	0.2323	0.2480	0.2208	0.2353	0.2295	0.2150	0.2135	0.2036
Honduras	0.3369	0.3192	0.3246	0.3337	0.3315	0.3333	0.3357	0.3282
Hungary	0.3261	0.3514	0.3567	0.3548	0.3629	0.3658	0.3690	0.3772
Iceland	0.5382	0.5508	0.5508	0.5535	0.5317	0.5380	0.5376	0.5452
India	0.1966	0.2383	0.2403	0.2567	0.2445	0.2474	0.2494	0.2544
Indonesia	0.2393	0.2444	0.2499	0.2545	0.2580	0.2774	0.2818	0.2761
Iran, Islamic Rep.	0.2567	0.2827	0.2854	0.2992	0.2942	0.2949	0.2959	0.2998
Ireland	0.3884	0.4216	0.4217	0.4236	0.4209	0.4163	0.4124	0.4187
Israel	0.3700	0.3878	0.3910	0.3943	0.3902	0.3907	0.3878	0.3817
Italy	0.4313	0.4436	0.4470	0.4529	0.4556	0.4670	0.4526	0.4602
Jamaica	0.2923	0.3032	0.3095	0.3123	0.3051	0.3143	0.3165	0.3073
Japan	0.5471	0.5115	0.5021	0.5075	0.4980	0.4873	0.5265	0.5328
Jordan	0.2698	0.2809	0.2837	0.2804	0.2908	0.2973	0.2905	0.2859
Kazakhstan	0.2645	0.2697	0.2864	0.2841	0.2912	0.2983	0.3004	0.3100
Kenya	0.2436	0.3040	0.3029	0.3125	0.2996	0.2892	0.3122	0.3102
Korea, Rep.	0.3576	0.3978	0.4025	0.4020	0.4042	0.4066	0.4147	0.4194
Kuwait	0.3206	0.3266	0.3371	0.3335	0.3308	0.3203	0.3237	0.3300
Kyrgyz Republic	0.3667	0.3720	0.3742	0.3959	0.3854	0.4097	0.3927	0.3989
Lao PDR	0.2318	0.2481	0.2601	0.2595	0.2606	0.2705	0.2647	0.2636
Latvia	0.4147	0.4514	0.4300	0.4425	0.4480	0.4573	0.4385	0.4397
Lebanon		0.2887	0.2991	0.3056	0.2983	0.3047	0.3053	0.3070
Lesotho	0.2910	0.2973	0.2861	0.3076	0.2869	0.2872	0.2831	0.2807
Lithuania	0.3140	0.3446	0.3460	0.3505	0.3505	0.3470	0.3736	0.3896
Luxembourg	0.4450	0.4139	0.4009	0.4223	0.4113	0.3758	0.3829	0.3924
Macedonia, FYR		0.3233	0.3327	0.3055	0.3146	0.3297	0.3574	0.3473
Madagascar	0.2576	0.2713	0.2800	0.2793	0.2868	0.2854	0.2813	0.2817
Malawi	0.2167	0.2052	0.2224	0.2472	0.2387	0.2497	0.2532	0.2500
Malaysia	0.3143	0.3336	0.3373	0.3449	0.3432	0.3422	0.3446	0.3464
Maldives	0.2571	0.2599	0.2552	0.2482	0.2547	0.2500	0.2553	0.2458
Mali	0.0870	0.0994	0.1049	0.1076	0.1045	0.1138	0.1055	0.1103
Mauritius	0.3094	0.3200	0.3215	0.3315	0.3274	0.3233	0.3246	0.3239
Mexico	0.3553	0.3639	0.3692	0.3646	0.3749	0.3657	0.3727	0.3703
Moldova	0.2460	0.2773	0.2787	0.2914	0.2892	0.2743	0.2771	0.2768
Mongolia	0.2180	0.2660	0.2664	0.2715	0.2598	0.2513	0.2366	0.2474

(continued)

Country	2000	2005	2006	2007	2008	2009	2010	2011
Morocco	0.2121	0.2565	0.2553	0.2655	0.2590	0.2750	0.2798	0.2637
Mozambique	0.2253	0.2314	0.2355	0.2331	0.2370	0.2421	0.2554	0.2635
Namibia	0.4058	0.4257	0.4209	0.4319	0.4290	0.4229	0.4216	0.4242
Nepal	0.3102	0.3431	0.3404	0.3478	0.3603	0.3477	0.3670	0.3617
Netherlands	0.3810	0.3984	0.4063	0.4220	0.4160	0.4082	0.4145	0.4329
New Zealand	0.4999	0.4985	0.4969	0.4993	0.4978	0.5150	0.5187	0.5233
Nicaragua	0.2410	0.2838	0.2814	0.2885	0.2918	0.2872	0.2974	0.2948
Niger	0.1549	0.1937	0.1907	0.1991	0.2080	0.2114	0.2163	0.2196
Norway	0.5633	0.5854	0.5865	0.5910	0.5889	0.5807	0.5767	0.5828
Oman		0.2604	0.2659	0.2586	0.2686	0.2703	0.2687	0.2723
Pakistan	0.2013	0.2402	0.2398	0.2408	0.2355	0.2397	0.2480	0.2427
Panama	0.4263	0.4330	0.4275	0.4409	0.4591	0.4582	0.4425	0.4123
Paraguay	0.4117	0.4275	0.4301	0.4295	0.4313	0.4288	0.4314	0.4324
Peru	0.4134	0.4066	0.4151	0.3978	0.3952	0.3956	0.3913	0.3925
Philippines	0.3187	0.3195	0.3227	0.3192	0.3271	0.3232	0.3153	0.3166
Poland	0.3272	0.3403	0.3435	0.3507	0.3555	0.3570	0.3569	0.3625
Portugal	0.3774	0.3547	0.3796	0.3830	0.3819	0.3893	0.4150	0.4076
Romania	0.3276	0.3523	0.3573	0.3622	0.3713	0.3708	0.3886	0.3825
Russian Federation	0.3055	0.3190	0.3188	0.3244	0.3387	0.3366	0.3402	0.3346
Rwanda	0.1348	0.2352	0.2368	0.2471	0.2558	0.2538	0.2544	0.2622
Saudi Arabia	0.2936	0.3141	0.3131	0.3122	0.3172	0.3071	0.3056	0.3153
Senegal	0.1429	0.1887	0.1834	0.1846	0.1879	0.1922	0.1975	0.1973
Sierra Leone	0.1550	0.1917	0.1941	0.1855	0.1952	0.1936	0.1933	0.1966
Singapore	0.3532	0.3935	0.3946	0.4152	0.4138	0.4124	0.4105	0.4119
Slovak Republic	0.3518	0.3653	0.3706	0.3828	0.3766	0.3702	0.3782	0.3843
Slovenia	0.3945	0.4180	0.4242	0.4308	0.4360	0.4342	0.4332	0.4299
Solomon Islands	0.1855	0.1631	0.2004	0.1887	0.1989	0.1950	0.1951	0.2003
South Africa	0.2490	0.2366	0.2400	0.2349	0.2344	0.2380	0.2396	0.2436
Spain	0.3825	0.3944	0.3995	0.4026	0.4061	0.4138	0.4219	0.4215
Sri Lanka	0.3407	0.3562	0.3716	0.3656	0.3659	0.3654	0.3901	0.3599
St. Vincent and the Grenadines							0.2893	0.2954
Sudan	0.1929	0.1581	0.1754	0.1605	0.1730	0.2121	0.2527	0.2480
Swaziland	0.2945	0.3109	0.3031	0.3135	0.3015	0.2849	0.2877	0.2954
Sweden	0.4917	0.4952	0.4914	0.5055	0.5074	0.5026	0.4969	0.4987
Switzerland	0.4932	0.5072	0.5000	0.5039	0.4950	0.5027	0.5106	0.5080
Syrian Arab Republic	0.2263	0.2343	0.2431	0.2558	0.2600	0.2630	0.2428	0.2496
Tajikistan	0.3483	0.3510	0.3728	0.3957	0.4006	0.3996	0.4022	0.4187
Tanzania	0.2703	0.2371	0.2252	0.2603	0.2715	0.2696	0.2636	0.2578
Thailand	0.2673	0.2911	0.2961	0.3138	0.3027	0.3048	0.3049	0.3084
Togo	0.2294	0.1986	0.2015	0.2228	0.2660	0.2764	0.2773	0.2789

(continued)

Country	2000	2005	2006	2007	2008	2009	2010	2011
Trinidad and Tobago	0.2850	0.2938	0.3007	0.2789	0.2745	0.2826	0.2814	0.2829
Tunisia	0.2S69	0.2757	0.2798	0.2779	0.2785	0.2878	0.2934	0.2867
Turkey	0.3011	0.3104	0.3178	0.3074	0.3107	0.3071	0.3190	0.3228
Uganda	0.2308	0.2533	0.2522	0.2539	0.2703	0.2676	0.2705	0.2749
Ukraine	0.2792	0.3122	0.3150	0.3175	0.3206	0.3066	0.3107	0.3093
United Kingdom	0.3576	0.3671	0.3618	0.3697	0.3764	0.3717	0.3684	0.3717
United States	0.4950	0.4385	0.4402	0.4317	0.4194	0.4211	0.4265	0.4203
Uruguay	0.4578	0.4590	0.4229	0.4620	0.4225	0.4395	0.4686	0.4476
Uzbekistan		0.2388	0.2450	0.2426	0.2591	0.2606	0.2654	0.2693
Vanuatu							0.2846	0.2871
Venezuela, RB	0.3911	0.3988	0.4087	0.4178	0.4248	0.4176	0.4160	0.4118
Vietnam	0.3016	0.2868	0.2934	0.2958	0.2904	0.3016	0.2927	0.2970
Yemen, Rep.	0.0875	0.0926	0.1123	0.1049	0.1266	0.1341	0.1420	0.1499
Zambia	0.2776	0.2884	0.2956	0.2748	0.2903	0.3123	0.3141	0.3148
Zimbabwe	0.1901	0.1816	0.2032	0.2010	0.2116	0.2159	0.2205	0.2291

Source: EIIW calculations based on data of World Bank, World Development Indicators and UN Comtrade, DESA/UNSD, UNDP Human Development Index

Appendix G: Comparison Between EIIW-vita Global Sustainability Indicator with Three Inputs and Extended Global Sustainability Indicator with Four Inputs Including Water Productivity, 2011

Country	EIIW-vita GSI (three inputs)	Ranking	Extended EIIW-vita GSI (four inputs)	Ranking
Albania	0.1925	10	0.1336	12
Algeria	−0.0593	87	−0.0516	92
Angola	0.0352	53	0.0358	49
Argentina	−0.0294	74	−0.0333	82
Armenia	−0.0079	68	−0.0177	71
Australia	−0.1004	111	−0.0780	112
Austria	0.1260	21	0.1066	21
Azerbaijan	−0.0790	100	−0.0710	108
Bahrain	−0.1175	122	−0.0889	120
Bangladesh	−0.0693	95	−0.0637	104
Belarus	−0.0958	109	−0.0815	114
Belgium	−0.0673	93	−0.0453	87
Belize	0.0344	54	0.0157	59
Benin	−0.1203	124	−0.0915	123
Bolivia	−0.0050	67	−0.0144	70

(continued)

Country	EIIW-vita GSI (three inputs)	Ranking	Extended EIIW-vita GSI (four inputs)	Ranking
Botswana	−0.0813	104	−0.0553	96
Brazil	0.1543	16	0.1085	20
Brunei Darussalam	−0.1309	132	−0.0813	113
Bulgaria	−0.0808	103	−0.0715	109
Burkina Faso	0.0967	31	0.0621	36
Burundi	0.0568	47	0.0316	53
Cambodia	−0.1125	119	−0.0956	130
Cameroon	0.1066	28	0.0731	33
Canada	0.0630	44	0.0421	46
Chile	−0.0028	66	−0.0108	67
China	0.1174	23	0.0777	29
Colombia	0.1338	19	0.0921	25
Congo, Rep.	−0.1562	138	−0.0824	116
Costa Rica	0.1848	12	0.1288	13
Cote d'Ivoire	−0.0317	76	−0.0328	81
Croatia	0.0325	55	0.0315	54
Cyprus	−0.1210	125	−0.0751	111
Czech Republic	−0.0768	99	−0.0462	88
Denmark	0.0407	51	0.1233	16
Dominican Republic	−0.1080	114	−0.0895	122
Ecuador	0.0401	52	0.0186	58
Egypt, Arab Rep.	−0.1098	117	−0.0942	128
El Salvador	0.0782	40	0.0499	43
Eritrea	−0.1514	137	−0.1254	140
Estonia	−0.0693	96	−0.0621	103
Ethiopia	0.2356	3	0.1654	5
Fiji	−0.1091	115	−0.0841	118
Finland	−0.0006	64	0.0215	57
France	−0.0805	102	−0.0538	95
Gabon	0.0275	58	0.0296	55
The Gambia	0.0913	34	0.0589	38
Georgia	0.1279	20	0.0849	26
Germany	0.2759	1	0.2197	1
Ghana	0.0634	43	0.0398	48
Greece	−0.1144	121	−0.0919	125
Guatemala	0.0819	39	0.0521	41
Guinea	−0.0018	65	−0.0131	69
Guinea-Bissau	0.1702	14	0.1164	18
Guyana	0.0059	60	−0.0078	64
Haiti	−0.0469	83	−0.0465	89
Honduras	0.0315	56	0.0139	60
Hungary	−0.0616	88	−0.0532	94

(continued)

Country	EIIW-vita GSI (three inputs)	Ranking	Extended EIIW-vita GSI (four inputs)	Ranking
Iceland	0.1924	11	0.1592	8
India	−0.0383	79	−0.0406	85
Indonesia	−0.0649	90	−0.0601	101
Iran, Islamic Rep.	−0.1075	113	−0.0923	127
Ireland	−0.0705	98	0.0056	62
Israel	−0.1247	127	−0.0824	115
Italy	0.0464	50	0.0329	51
Jamaica	−0.1125	120	−0.0922	126
Japan	0.1647	15	0.1248	14
Jordan	−0.1262	130	−0.1021	132
Kazakhstan	−0.1250	129	−0.1051	135
Kenya	0.1114	27	0.0736	31
Korea, Rep.	−0.0582	86	−0.0450	86
Kuwait	−0.1001	110	−0.0585	98
Kyrgyz Republic	0.1828	13	0.1248	15
Lao PDR	0.0033	61	−0.0096	65
Latvia	0.0744	41	0.0540	39
Lebanon	−0.1250	128	−0.0998	131
Lesotho	0.1115	26	0.0810	28
Lithuania	−0.0308	75	−0.0322	80
Luxembourg	−0.0821	105	0.1118	19
Macedonia, FYR	−0.0334	77	−0.0355	84
Madagascar	0.0834	37	0.0503	42
Malawi	0.0999	30	0.0637	35
Malaysia	−0.0682	94	−0.0597	100
Maldives	−0.1693	142	−0.0659	105
Mali	−0.1384	135	−0.1158	138
Mauritius	−0.0802	101	−0.0695	107
Mexico	−0.0293	73	−0.0312	79
Moldova	−0.0954	108	−0.0834	117
Mongolia	−0.1583	139	−0.1285	141
Morocco	−0.0545	85	−0.0516	91
Mozambique	0.2050	7	0.1449	9
Namibia	0.2234	5	0.1634	6
Nepal	0.2653	2	0.1869	3
Netherlands	−0.0443	82	−0.0282	76
New Zealand	0.1387	18	0.0984	23
Nicaragua	0.0005	63	−0.0104	66
Niger	0.1442	17	0.0962	24
Norway	0.2226	6	0.1836	4
Oman	−0.1603	140	−0.1238	139
Pakistan	−0.0186	71	−0.0261	75
Panama	0.0567	48	0.0453	45

(continued)

Country	EIIW-vita GSI (three inputs)	Ranking	Extended EIIW-vita GSI (four inputs)	Ranking
Paraguay	0.1999	8	0.1439	10
Peru	0.0599	45	0.0342	50
Philippines	−0.0109	69	−0.0201	74
Poland	−0.0880	107	−0.0694	106
Portugal	0.0063	59	−0.0015	63
Romania	−0.0161	70	−0.0200	73
Russian Federation	−0.0858	106	−0.0729	110
Rwanda	0.0954	32	0.0665	34
Saudi Arabia	−0.1393	136	−0.1125	137
Senegal	−0.0645	89	−0.0594	99
Sierra Leone	0.0571	46	0.0317	52
Singapore	−0.0422	80	0.2061	2
Slovak Republic	−0.0654	92	−0.0306	78
Slovenia	−0.0242	72	−0.0195	72
Solomon Islands	−0.1095	116	−0.0844	119
South Africa	−0.1318	133	−0.1048	134
Spain	−0.0351	78	−0.0290	77
Sri Lanka	0.0288	57	0.0100	61
St. Vincent and the Grenadii	−0.1263	131	−0.0915	124
Sudan	0.0880	36	0.0539	40
Swaziland	0.0687	42	0.0399	47
Sweden	0.0934	33	0.1001	22
Switzerland	0.1131	24	0.1169	17
Syrian Arab Republic	−0.1329	134	−0.1115	136
Tajikistan	0.2305	4	0.1606	7
Tanzania	0.0495	49	0.0259	56
Thailand	−0.0651	91	−0.0602	102
Togo	0.1228	22	0.0B38	27
Trinidad and Tobago	−0.1942	143	−0.1367	143
Tunisia	−0.1247	126	−0.1022	133
Turkey	−0.0534	B4	−0.0484	90
Uganda	0.1039	29	0.0775	30
Ukraine	−0.1105	118	−0.0946	129
United Kingdom	−0.1196	123	−0.0531	93
United States	−0.0694	97	−0.0570	97
Uruguay	0.1122	25	0.0735	32
Uzbekistan	−0.1025	112	−0.0891	121
Vanuatu	−0.0429	81	−0.0344	83
Venezuela, RB	0.0885	35	0.0594	37
Vietnam	0.0010	62	−0.0114	68

(continued)

Country	EIIW-vita GSI (three inputs)	Ranking	Extended EIIW-vita GSI (four inputs)	Ranking
Yemen, Rep.	−0.1623	141	−0.1327	142
Zambia	0.1995	9	0.1389	11
Zimbabwe	0.0822	38	0.0496	44

Source: EIIW calculations based on data of World Bank, World Development Indicators and UN Comtrade, DESA/UNSD

Appendix H: Four Sub-indicators of the Extended EIIW-vita Global Sustainability Indicator, 2011

Country	Savings indicator	Ranking	RCA indicator	Ranking	Renewables indicator	Ranking	Water productivity Indicator	Ranking
Albania	−0.0494	110	−0.0137	49	0.6407	3	−0.0430	98
Algeria	0.1876	4	−0.0161	104	−0.3493	128	−0.0287	60
Angola	−42,306	140	−0.0139	66	0.3500	22	0.0377	21
Argentina	0.0347	46	−0.0241	122	−0.0988	73	−0.0448	106
Armenia	0.0136	65	−0.0137	46	−0.0238	63	−0.0471	123
Australia	−0.0018	76	−0.0413	132	−0.2581	102	−0.0109	44
Austria	0.0761	24	0.0043	16	0.2975	28	0.0484	18
Azerbaijan	0.0039	70	−0.0137	59	−0.2272	93	−0.0469	122
Bahrain	0.0207	58	−0.0141	78	−0.3591	140	−0.0030	33
Bangladesh	0.1466	7	−0.0152	98	−0.3393	125	−0.0469	120
Belarus	0.0861	19	−0.0192	114	−0.3544	132	−0.0383	83
Belgium	0.0575	35	0.0056	15	−0.2650	105	0.0206	27
Belize	0.0258	55	−0.0136	26	0.0909	53	−0.0403	88
Benin	0.0054	69	−0.0136	32	−0.3526	130	−0.0050	34
Bolivia	40,013	73	−0.0136	34	0.0000	61	−0.0428	96
Botswana	0.1290	9	−0.0137	57	−0.3591	139	0.0226	26
Brazil	−40,018	75	−0.0474	133	0.5121	13	−0.0287	61
Brunet Darussalam	0.0200	93	−0.0139	70	−0.3586	135	0.0673	12
Bulgaria	0.0543	38	−0.0164	106	−0.2804	114	−0.0436	100
Burkina Faso	0.0127	67	−0.0136	36	0.2909	34	−0.0414	91
Burundi	−0.1068	133	−0.0136	27	0.2909	31	−0.0441	102
Cambodia	−0.0293	103	−0.0136	41	−0.2945	117	−0.0449	107
Cameroon	−0.0506	114	−0.0143	81	0.3846	20	−0.0273	57
Canada	−40,023	77	−0.0727	140	0.2641	42	−0.0208	50

(continued)

Country	Savings indicator	Ranking	RCA indicator	Ranking	Renewables indicator	Ranking	Water productivity Indicator	Ranking
Chile	−0.0256	98	−0.0198	115	0.0369	60	−0.0347	74
China	0.2392	2	0.3016	3	−0.1887	83	−0.0414	90
Colombia	−0.0465	108	−0.0167	108	0.4645	14	−0.0330	67
Congo, Rep.	−47,074	143	−0.0139	65	0.2527	43	0.1389	7
Costa Rica	0.0163	63	−0.0151	95	0.5532	12	−0.0390	85
Cote d'Ivoire	−0.0234	94	−0.0139	69	−0.0577	66	−0.0363	77
Croatia	0.0297	51	−0.0168	109	0.0845	55	0.0288	24
Cyprus	−40,255	97	−0.0145	85	−0.3230	123	0.0624	15
Czech Republic	−0.0149	90	0.0602	7	−0.2756	109	0.0455	19
Denmark	0.0645	30	0.0140	12	0.0436	59	0.3709	3
Dominican Republic	−40,711	120	−0.0140	72	−0.2390	97	−0.0341	70
Ecuador	−40,831	123	−0.0147	89	0.2182	44	−0.0458	116
Egypt, Arab Rep.	−0.0453	106	−0.0200	116	−0.2640	103	−0.0475	128
El Salvador	−0.0515	116	−0.0143	82	0.3003	27	−0.0349	75
Eritrea	−0.0283	102	−0.0727	141	−0.3532	131	−0.0473	125
Estonia	0.0751	27	−0.0154	99	−0.2676	107	−0.0404	89
Ethiopia	0.0859	20	−0.0136	39	0.6345	7	−0.0453	112
Fiji	−0.1045	130	−0.0136	40	−0.2091	89	−0.0092	39
Finland	0.0364	44	−0.0076	19	−0.0304	64	0.0875	10
France	0.0160	64	−0.0148	90	−0.2428	98	0.0264	25
Gabon	−0.0060	80	−0.0147	88	0.1033	52	0.0356	23
The Gambia	−0.0033	78	−0.0136	24	0.2909	30	−0.0382	82
Georgia	−0.0174	91	−0.0138	62	0.4149	15	−0.0442	103
Germany	0.0550	37	0.9273	1	−0.1547	78	0.0512	17
Ghana	−41,104	134	−0.0154	100	0.3160	24	−0.0309	63
Greece	−0.1018	126	−0.0201	117	−0.2215	92	−0.0242	54
Guatemala	0.0498	113	−0.0146	87	0.3100	26	−0.0371	81
Guinea	0.2825	142	−0.0137	55	0.2909	38	−0.0472	124
Guinea-Bissau	0.2926	1	−0.0727	138	0.2909	41	−0.0451	109
Guyana	−40,597	118	−0.0136	29	0.0909	54	−0.0488	139
Haiti	0.0808	23	−0.0727	139	−0.1488	76	−0.0454	113
Honduras	0.0326	50	−0.0140	71	0.0758	56	−0.0388	84
Hungary	0.0483	41	0.0508	9	−0.2838	115	−0.0283	59
Iceland	0.0497	112	−0.0139	67	0.6408	2	0.0595	16
India	0.1339	8	−0.0638	137	−0.1849	82	−0.0476	129
Indonesia	0.0816	22	−0.0377	131	−0.2385	96	−0.0456	115
Iran, Islamic Rep.	0.0133	66	−0.0278	125	−0.3079	120	−0.0469	121
Ireland	−40,124	87	−0.0363	129	−0.1629	79	0.2340	5

(continued)

Country	Savings indicator	Ranking	RCA indicator	Ranking	Renewables indicator	Ranking	Water productivity Indicator	Ranking
Israel	−40,057	79	−0.0138	63	−0.3546	134	0.0444	20
Italy	−40,014	74	0.2238	4	−0.0831	70	−0.0079	38
Jamaica	−40,466	109	−0.0137	44	−0.2772	110	−0.0315	64
Japan	0.0276	53	0.7033	2	−0.2369	95	0.0053	31
Jordan	−40,098	83	−0.0143	80	−0.3546	133	−0.0296	62
Kazakhstan	−40,921	124	−0.0148	91	−0.2681	108	−0.0452	111
Kenya	0.0348	45	−0.0141	76	0.3136	25	−0.0399	87
Korea, Rep.	0.1236	10	0.0464	10	−0.3447	126	−0.0052	35
Kuwait	0.0734	28	−0.0144	84	−0.3591	141	0.0660	14
Kyrgyz Republic	−0.0118	86	−0.0137	50	0.5737	11	−0.0491	142
Lao PDR	−0.0712	122	0.0119	14	0.0691	57	−0.0483	135
Latvia	0.0931	18	−0.0156	103	0.1458	49	−0.0072	37
Lebanon	−0.0506	115	−0.0144	83	−0.3099	122	−0.0243	55
Lesotho	0.0572	36	−0.0136	30	0.2909	32	−0.0104	42
Lithuania	0.0214	57	−0.0168	110	−0.0970	72	0.0365	78
Luxembourg	0.0440	42	−0.0228	119	−0.2676	106	0.6936	2
Macedonia, FYR	0.0643	31	−0.0141	77	−0.1505	77	−0.0418	92
Madagascar	−0.0270	100	−0.0137	56	0.2909	39	−0.0490	141
Malawi	0.0226	56	−0.0137	58	0.2909	40	−0.0450	108
Malaysia	0.1102	13	−0.0247	124	−0.2902	116	−0.0343	72
Maldives	−0.2262	139	−0.0727	143	−0.2091	91	0.2445	4
Mall	−0.0923	125	−0.0137	47	−0.3091	121	−0.0483	133
Mauritius	−0.0180	92	−0.0136	42	−0.2091	90	−0.0371	80
Mexico	0.0435	43	0.0688	6	−0.2003	84	−0.0366	79
Moldova	0.0260	54	−0.0139	68	−0.2983	118	−0.0474	126
Mongolia	−0.1021	127	−0.0136	35	−0.3591	137	−0.0393	86
Morocco	0.1075	14	−0.0151	93	−0.2560	101	−0.0427	94
Mozambique	−0.0112	85	0.0137	53	0.6397	5	−0.0354	76
Namibia	0.0614	33	0.0140	74	0.6227	9	−0.0163	47
Nepal	0.1696	5	−0.0137	54	0.6400	4	−0.0483	134
Netherlands	0.0655	29	0.0517	8	−0.2501	99	0.0200	28
New Zealand	0.0344	48	−0.0178	113	0.3995	16	−0.0226	51
Nicaragua	0.0342	49	−0.0137	45	−0.0190	62	−0..0431	99
Niger	0.1553	6	−0.0136	37	0.2909	35	−0.0477	131
Norway	0.0992	15	−0.0376	130	0.6062	10	0.0666	13
Oman	−0.1043	129	−0.0176	112	−0.3591	142	−0.0143	46
Pakistan	0.0184	60	−0.0146	86	−0.0597	67	−0.0486	138
Panama	0.0196	59	−0.0150	92	0.1654	48	0.0111	30
Paraguay	−0.0276	101	−0.0137	51	0.6409	1	−0.0241	53
Peru	−0.0130	88	−0.0168	111	0.2096	45	−0.0428	95

(continued)

Country	Savings indicator	Ranking	RCA indicator	Ranking	Renewables indicator	Ranking	Water productivity Indicator	Ranking
Philippines	0.0618	32	−0.0219	118	−0.0725	69	−0.0477	130
Poland	0.0174	61	−0.0029	18	−0.2786	111	−0.0138	45
Portugal	−0.0704	119	−0.0165	107	0.1056	51	−0.0250	56
Romania	0.0600	34	−0.0123	21	−0.0959	71	−0.0316	65
Russian Federation	0.0012	72	−0.0574	134	−0.2011	85	−0.0342	71
Rwanda	0.0088	68	−0.0136	31	0.2909	33	−0.0200	49
Saudi Arabia	−0.0260	99	−0.0329	128	−0.3591	143	−0.0320	66
Senegal	0.0756	26	−0.0138	61	−0.2552	100	−0.0444	104
Sierra Leone	−0.1060	132	−0.0136	22	0.2909	29	−0.0445	105
Singapore	0.2330	3	−0.0233	121	−0.3362	124	0.9509	1
Slovak Republic	0.0017	71	−0.0154	101	−0.1824	81	0.0736	11
Slovenia	0.0503	40	−0.0077	20	−0.1154	75	−0.0053	36
Solomon Islands	−0.1058	131	−0.0136	25	−0.2091	87	−0.0092	40
South Africa	−0.0444	105	−0.0012	17	−0.3496	129	−0.0239	52
Spain	0.0166	62	−0.0611	136	−0.0608	68	−0.0105	43
Sri Lanka	0.0518	39	−0.0151	94	0.0498	58	−0.0465	118
St. Vincent and the Grenadines	−0.1561	138	−0.0136	23	−0.2091	86	0.0129	29
Sudan	−0.0561	117	−0.0727	142	0.3928	18	−0.0484	137
Swaziland	−0.0711	121	−0.0137	48	0.2909	36	−0.0464	117
Sweden	0.0952	16	−0.0154	102	0.2004	46	0.1204	9
Switzerland	0.1217	11	0.0360	11	0.1814	47	0.1285	8
Syrian Arab Republic	−0.1033	128	−0.0161	105	−0.2792	113	−0.0474	127
Tajikistan	0.0760	25	−0.0136	33	0.6291	8	−0.0491	143
Tanzania	0.0285	52	−0.0141	75	0.1341	50	−0.0451	110
Thailand	0.1133	12	−0.0301	127	−0.2786	112	−0.0455	114
Togo	−0.0143	89	−0.0137	60	0.3963	17	−0.0331	68
Trinidad and Tobago	−0.2357	141	0.0123	13	−0.3591	136	0.0358	22
Tunisia	−0.0098	84	−0.0151	96	−0.3490	127	−0.0347	73
Turkey	−0.0245	95	−0.0300	126	−0.1058	74	−0.0331	69
Uganda	0.0344	47	−0.0137	52	0.2909	37	−0.0016	32
Ukraine	−0.0069	81	−0.0229	120	−0.3016	119	−0.0468	119
United Kingdom	−0.0350	104	−0.0590	135	−0.2648	104	0.1465	6
United States	−0.0495	111	0.0781	5	−0.2368	94	−0.0200	48
Uruguay	−0.0094	82	−0.0140	73	0.3599	21	−0.0423	93
Uzbekistan	−0.1292	137	−0.0138	64	−0.1644	80	−0.0490	140

(continued)

Country	Savings indicator	Ranking	RCA indicator	Ranking	Renewables indicator	Ranking	Water productivity Indicator	Ranking
Vanuatu	0.0941	17	−0.0136	28	−0.2091	88	−0.0092	41
Venezuela, RB	−0.0457	107	−0.0152	97	0.3264	23	−0.0280	58
Vietnam	0.0842	21	−0.0244	123	−0.0570	65	−0.0484	136
Yemen, Rep.	−0.1141	135	−0.0136	38	−0.3591	138	−0.0439	101
Zambia	−0.0246	96	−0.0142	79	0.6374	6	−0.0430	97
Zimbabwe	−0.1248	136	−0.0136	43	0.3849	19	−0.0480	132

Source: EIIW calculations based on data of World Bank, World Development Indicators and UN Comtrade, DESA/UNSD

Appendix I: Total Primary Supply of Renewable Energy in Selected Asian Countries

(a) Brunei Darussalam (in kt oil equivalent)

Type of renewable energy	2002	2003	2004	2005	2006	2007	2008	2009	2010	2011	2012
Nuclear	0	0	0	0	0	0	0	0	0	0	0
Hydro	0	0	0	0	0	0	0	0	0	0	0
Geothermal	0	0	0	0	0	0	0	0	0	0	0
Solar/wind/other	0	0	0	0	0	0	0	0	0	0.172	0.172
Biofuels	0	0	0	0	0	0	0	0	0	0	0

Data Source: International Energy Agency, Energy Balance of Non-OECD countries, 2014

(b) Cambodia (in kt oil equivalent)

Type of renewable energy	2002	2003	2004	2005	2006	2007	2008	2009	2010	2011	2012
Nuclear	0	0	0	0	0	0	0	0	0	0	0
Hydro	2.494	3.526	2.322	3.784	4.128	4.042	3.698	3.182	2.236	3.870	44.462
Geothermal	0	0	0	0	0	0	0	0	0	0	0
Solar/wind/other	0.086	0.086	0.086	0.086	0.172	0.172	0.172	0.172	0.258	0.258	0.258
Biofuels	0	0	0	0	0	0	0	0	0	0	0

Data Source: International Energy Agency, Energy Balance of Non-OECD countries, 2014

(c) Indonesia (in kt oil equivalent)

Type of renewable energy	2002	2003	2004	2005	2006	2007	2008	2009	2010	2011	2012
Nuclear	0	0	0	0	0	0	0	0	0	0	0
Hydro	854.238	782.514	831.964	922.350	827.578	970.596	991.408	979.024	1501.216	1068.034	1100.714
Geothermal	10,725.368	10,821.652	11,444.06	11,354.653	11,447.499	12,071.627	14,286.162	15,981.451	16,088.052	16,112.123	16,191.213
Solar/wind/other	0	0	0	0	0	0	0	0.344	0.430	0.516	0.688
Biofuels	0	0	0	0	4.395	17.980	18.219	49.383	171.396	279.507	523.856

Data Source: International Energy Agency, Energy Balance of Non-OECD countries, 2014

(d) Malaysia (in kt oil equivalent)

Type of renewable energy	2002	2003	2004	2005	2006	2007	2008	2009	2010	2011	2012
Nuclear	0	0	0	0	0	0	0	0	0	0	0
Hydro	456.144	434.73	501.38	446.426	554.012	558.14	641.646	598.474	556.592	655.578	778.816
Geothermal	0	0	0	0	0	0	0	0	0	0	0
Solar/wind/other	0.086	0.086	0.086	0.086	0.086	0.086	0	0	0	0	4.042
Biofuels	0	0	0	0	6.153	30.763	42.19	4.395	4.395	23.732	116.022

Data Source: International Energy Agency, Energy Balance of Non-OECD countries, 2014

(e) Myanmar (in kt oil equivalent)

Type of renewable energy	2002	2003	2004	2005	2006	2007	2008	2009	2010	2011	2012
Nuclear	0	0	0	0	0	0	0	0	0	0	0
Hydro	181.546	178.450	207.088	257.742	285.950	311.234	350.106	452.016	439.030	646.548	667.876
Geothermal	0	0	0	0	0	0	0	0	0	0	0
Solar/wind/other	0	0	0	0	0	0	0	0	0	0	0
Biofuels	0	0	0	0	0	0	0	0	0	0	0

Data Source: International Energy Agency, Energy Balance of Non-OECD countries, 2014

(f) Philippines (in kt oil equivalent)

Type of renewable energy	2002	2003	2004	2005	2006	2007	2008	2009	2010	2011	2012
Nuclear	0	0	0	0	0	0	0	0	0	0	0
Hydro	604.838	676.820	738.998	721.282	854.754	736.418	846.498	841.768	671.058	834.028	881.672
Geothermal	8804.843	8443.777	8839.230	8512.551	8996.551	8781.631	9218.349	8875.336	8535.763	8546.939	8811.720
Solar/wind/other	0	0	0	1.634	4.644	5.074	5.332	5.590	5.418	7.654	6.536
Biofuels	0	0	0	1.417	2.359	33.198	56.98	204.804	213.562	222.536	293.583

Data Source: International Energy Agency, Energy Balance of Non-OECD countries, 2014

(g) Singapore (in kt oil equivalent)

Type of renewable energy	2002	2003	2004	2005	2006	2007	2008	2009	2010	2011	2012
Nuclear	0	0	0	0	0	0	0	0	0	0	0
Hydro	0	0	0	0	0	0	0	0	0	0	0
Geothermal	0	0	0	0	0	0	0	0	0	0	0
Solar/wind/other	0	0	0	0	0	0	0.086	0.258	0.430	0.688	1.118
Biofuels	0	0	0	0	0	0	0	0	0	0	0

Data Source: International Energy Agency, Energy Balance of Non-OECD countries, 2014

(h) Thailand (in kt oil equivalent)

Type of renewable energy	2002	2003	2004	2005	2006	2007	2008	2009	2010	2011	2012
Nuclear	0	0	0	0	0	0	0	0	0	0	0
Hydro	642.506	627.714	519.440	498.628	698.750	697.804	611.718	614.728	476.182	702.018	752.844
Geothermal	1.719	1.719	1.719	1.719	1.719	1.719	0.860	0.860	1.719	0.860	0.860
Solar/wind/other	0	0	0	0	0.086	0.086	0.258	0.860	1.720	8.600	54.524
Biofuels	0	0	3.201	33.926	64.011	141.072	478.705	633.954	654.409	706.745	926.923

Data Source: International Energy Agency, Energy Balance of Non-OECD countries, 2014

(i) Vietnam (in kt oil equivalent)

Type of renewable energy	2002	2003	2004	2005	2006	2007	2008	2009	2010	2011	2012
Nuclear	0	0	0	0	0	0	0	0	0	0	0
Hydro	1565.028	1632.796	1532.348	1457.270	1755.088	1981.010	2234.796	2578.366	2369.300	3519.378	4591.024
Geothermal	0	0	0	0	0	0	0	0	0	0	0
Solar/wind/ other	0	0	0	0	0	0	0.086	0.860	4.300	7.482	7.482
Biofuels	0	0	0	0	0	0	0	0	0	0	0

Data Source: International Energy Agency, Energy Balance of Non-OECD countries, 2014

(j) China including Hong Kong (in kt oil equivalent)

Type of renewable energy	2002	2003	2004	2005	2006	2007	2008	2009	2010	2011	2012
Nuclear	6548.248	11,295.188	13,152.527	13,835.055	14,292.418	16191.455	17,823.891	18,277.345	19,253.576	22,503.333	25,381.467
Hydro	24,765.764	24,396.566	30,404.784	34,143.462	37,477.596	41,732.704	50326.082	52,945.04	61,186.592	59,171.87	74,201.402
Geothermal	1838.187	1974.016	2109.846	2344.538	2624.579	2886.591	3181.246	3459.567	3730.175	4000.760	4271.344
Solar/wind/other	1510.239	1883.637	2267.637	2937.327	3781.32	4733.002	6693.766	9517.542	12,296.136	17,140.528	21,683.067
Biofuels	0	0	0	0	1007.93	932.836	1174.836	1211.235	1161.622	1209.2	1209.2

Data Source: International Energy Agency, Energy Balance of Non-OECD countries, 2014

(k) Korea (in kt oil equivalent)

Type of renewable energy	2002	2003	2004	2005	2006	2007	2008	2009	2010	2011	2012
Nuclear	31,038.964	33,793.309	34,065.121	38,251.497	38,764.891	37,250.248	39,340.57	38,510.018	38,725.018	40,321.752	39,176.127
Hydro	278.038	421.572	372.380	315.878	298.248	312.266	264.020	241.918	316.652	395.428	341.334
Geothermal	0.119	0.382	1.337	2.555	6.209	11.104	15.713	22.113	33.432	47.832	65.264
Solar/wind/other	36.661	35.745	41.032	47.192	56.824	68.393	91.761	145.902	182.849	205.888	233.504
Biofuels	0.913	1.825	4.563	10.952	40.156	79.399	149.671	213.555	300.255	282.915	302.993

Data Source: International Energy Agency, Energy Balance of Non-OECD countries, 2014

(l) Japan (in kt oil equivalent)

Type of renewable energy	2002	2003	2004	2005	2006	2007	2008	2009	2010	2011	2012
Nuclear	76,903.285	62,548.842	73,606.097	79,421.000	79,074.655	68,756.218	67,269.721	72,904.545	75,114.485	26,519.533	4153.800
Hydro	7084.508	8136.202	8089.418	6576.420	7527.064	6364.774	6574.528	6613.056	7070.232	7154.942	6491.796
Geothermal	3128.495	3224.015	3122.167	2990.015	2854.496	2820.562	2547.853	2690.369	2439.748	2480.654	2419.259
Solar/wind/other	832.955	785.053	793.694	846.989	883.411	908.434	934.237	991.210	1082.401	1222.856	1362.077
Biofuels	0	0	0	0	0	0	0	0	0	0	0

Data Source: International Energy Agency, Energy Balance of Non-OECD countries, 2014

(m) Pakistan (in kt oil equivalent)

Type of renewable energy	2002	2003	2004	2005	2006	2007	2008	2009	2010	2011	2012
Nuclear	453.455	458.667	728.394	647.345	596.267	801.885	421.661	754.194	891.273	1372.091	1186.539
Hydro	1922.186	2317.184	2207.706	2654.132	2747.958	2468.802	2389.424	2415.998	2735.746	2452.462	2567.702
Geothermal	0	0	0	0	0	0	0	0	0	0	0
Solar/wind/other	0	0	0	0	0	0	0	0	0	0	0
Biofuels	0	0	0	0	0	0	0	0	0	0	0

Data Source: International Energy Agency, Energy Balance of Non-OECD countries, 2014

(n) India (in kt oil equivalent)

Type of renewable energy	2002	2003	2004	2005	2006	2007	2008	2009	2010	2011	2012
Nuclear	4633.576	4433.17	4514.739	4899.915	4419.097	3890.067	4856.655	6845.079	8414.188	8566.382	5075.303
Hydro	6948.972	7784.032	9280.260	10351.992	10996.304	10045.058	9729.696	10584.02	12348.052	10821.122	6337.942
Geothermal	0	0	0	0	0	0	0	0	0	0	0
Solar/wind/other	360.412	450.659	630.784	955.551	1185.329	1367.175	1846.103	1992.249	2532.469	3078.669	231.44
Biofuels	86.414	91.535	96.656	101.777	101.777	101.777	142.104	74.3	66.036	232.216	209.401

Data Source: International Energy Agency, Energy Balance of Non-OECD countries, 2014

Appendix J: EIIW-vita Global Sustainability Indicator, Example

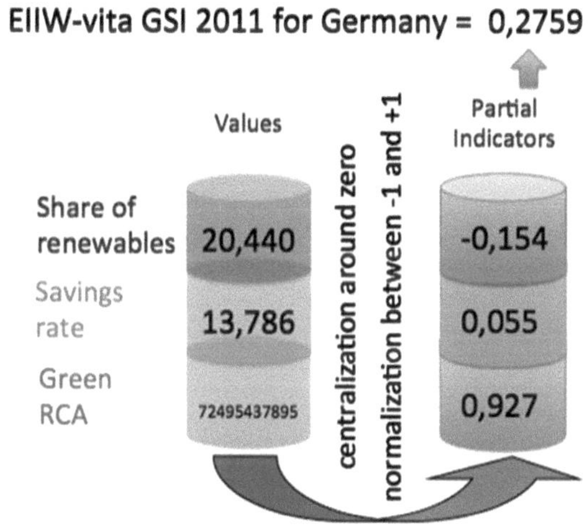

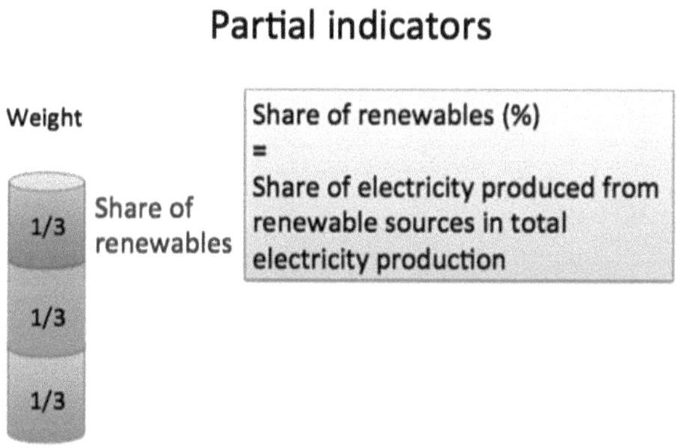

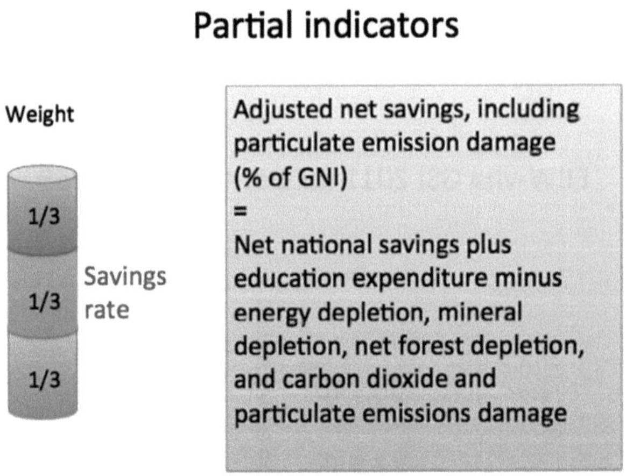
Partial indicators
Weight
1/3
1/3
1/3
Savings rate
Adjusted net savings, including particulate emission damage (% of GNI)
=
Net national savings plus education expenditure minus energy depletion, mineral depletion, net forest depletion, and carbon dioxide and particulate emissions damage

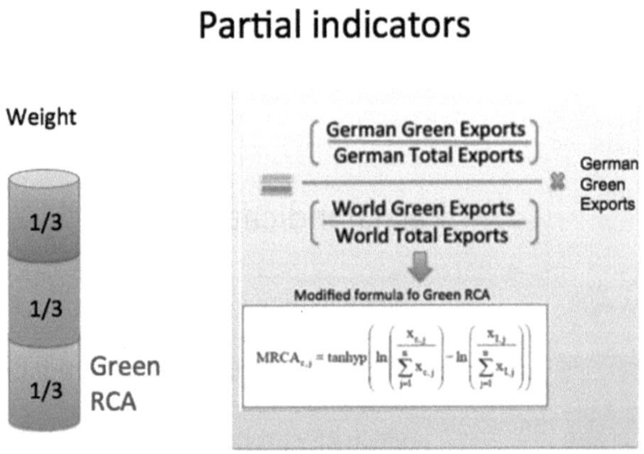
Partial indicators
Weight
1/3
1/3
1/3
Green RCA
German Green Exports
German Total Exports
World Green Exports
World Total Exports
German Green Exports
Modified formula fo Green RCA